This volume is one of a series of textbooks on psychology issued under the consulting editorship of Gregory A. Kimble

H. D. Kimmel, Ph.D., University of Southern California, is Professor and Chairman of the Department of Psychology, University of South Florida. He previously taught at the University of Florida and at Ohio University. He has been a visiting professor at Duke University, University of Puerto Rico, and University of Southern California. Dr. Kimmel is a regular consulting editor for the *Journal of Experimental Psychology* and is a Fellow in the Division of Experimental Psychology, American Psychological Association.

EXPERIMENTAL PRINCIPLES AND DESIGN IN PSYCHOLOGY

H. D. KIMMEL
UNIVERSITY OF SOUTH FLORIDA

THE RONALD PRESS COMPANY • NEW YORK

VR

Library of Congress Catalog Card Number: 70–137780

PRINTED IN THE UNITED STATES OF AMERICA

Preface

This text for advanced courses in experimental psychology describes the major methods that have been developed for planning and conducting experiments on behavior and analyzing their results. It deals with (1) experimental principles in relation to the general background in which experiments are done and what they achieve; (2) quantitative matters dealing with measurement in psychology and psychological scaling; (3) experimental design principles and methods.

Much of the emphasis is on what has come to be called experimental design. Although experimental design is often presented as a subdivision of statistics, an effort is made in this book to communicate the essential principles of design in as non-statistical a manner as possible. While giving the necessary statistical principles, the book endeavors to present them in a logical, as distinct from a purely mathematical–statistical, manner.

The first part of the text—Chapters 2 through 5—deals with the logical and methodological aspects of experimentation in psychology. Chapters 6 and 7 examine the ways in which statistical inference is used in planning data collection and analysis. Chapters 8 through 11 present a variety of experimental design methods that are used in psychological research. Chapter 12 concludes with some general principles of psychological experimentation.

This text is used by the author in a one-semester course, integrating the logical-conceptual aspects of psychological experimentation with its statistical-design features. He bases this preference on the belief that sufficient interrelationships can be identified to justify a unifying effort of this type. Interweaving the logical-conceptual and statistical-design principles of psychological experimentation has the highly desirable effect of keeping the psychological experimenter fully in the driver's seat in the planning, conduct, analysis, and interpretation of his research.

The organization of the text, however, also lends itself to a two-quarter sequence. Chapters 1 through 5 plus Chapter 12 deal exclusively with principles of psychological experimentation of a non-statistical type. Chapters 6 through 11 concentrate especially on experimental design and analysis.

Credit should be given to those of the author's own teachers who most influenced his thinking in matters of psychological experimentation—William W. Grings, David A. Grant, and J. P. Guilford. Those familiar with the work of these men will not fail to recognize their contributions. The author also thanks Gregory A. Kimble for his many helpful editorial recommendations. Finally, gratitude is due to the author's wife, who painstakingly read and reread every version of the manuscript until it met her standards.

H. D. Kimmel

Tampa, Florida
September, 1970

Contents

EXPERIMENTAL PRINCIPLES AND DESIGN IN PSYCHOLOGY

1

The Science of Psychology

The basic aim of science is to increase man's knowledge of himself and the world in which he lives. To some extent, therefore, scientists, philosphers, and even artists share a common concern in understanding the nature of things. It may be argued that the *methods* used to acquire knowledge provide the most effective means for identifying the differences among individuals engaged in scientific, philosophical, and artistic endeavors (and most of this book is about methods). It may also be fruitful, however, to give some initial consideration to the possibility that the very meaning of the word "knowledge" determines the manner in which it is sought.

SCIENTIFIC KNOWLEDGE

Although there is a peculiar redundancy in the phrase "scientific knowledge," since the word "science" came originally from the Latin word *scientia,* meaning knowledge, it has come nevertheless to mean something rather different from the ratiocination, insight, intuition, and, sometimes even, feeling which constitute knowledge for philosophers and artists. For the scientist, knowledge is *public,* in the sense that it is communicable in some fashion to others; *objective,* in that it fairly encompasses all extant facts; *verifiable,* because it implies feasible concrete observations; and *relational,* in that it identifies the ordered, systematic, and hopefully causal relations among events.

It may be important, also, to note that scientific knowledge is *conceptual,* a feature it shares with other, nonscientific types of knowledge. No matter how communicable, objective, and verifiable by means of empirical observations it may be, knowledge itself is originally and finally (and, to some, regrettably) conceived within the thoughts of men. The relationships among the properties of objects and events, which constitute scientific knowledge, themselves have no objective manifestation, except in the empirical observations they encompass and in the scientists' public communications regarding them. Although knowledge is about the objective world, it is, itself, conceptual.

That the scientist deals in the realm of ideas is to some so obvious as to preclude even mentioning it. The manner in which these ideas emerge from observation tends to place strong emphasis on the objective details of scientific procedures, and rightly so; but it may be useful to point out that these objective details are only relevant because of their contribution to the growing understanding of the workings of nature which scientists call knowledge. So, the scientist lives continuously with a conundrum. He must devise his investigations of nature at a public level, so that the conceptual constructions he employs to interrelate and organize his objective results will be both communicable to others and verifiable concretely. But he must avoid the snare of *scientism,* in which he becomes a tool of his own tools and in which he loses contact with ideas and knowledge and makes observations more or less purposelessly, somewhat in the fashion of a randomly rotating camera which takes photographs aimlessly.

Psychological science is no different in this regard from biological or physical science. The only differences, in fact, among these major subdivisions of science are to be found in their somewhat arbitrary selections of subject matter for study. The physical scientist usually deals with nonliving matter and energy. The biological scientist studies the basic structural, physiological, and biochemical properties of living things, and of life itself. The psychological scientist concentrates on the observable behavior of intact organisms, although he often employs the word "behavior" to include internal events which are picked up and transmitted to external recording devices. The borderlines among these different types of scientific activities are usually blurred and ambiguous, rather than clear and distinct. Many individuals trained as biologists are engaged in research involving the overt behavior of intact organisms; others, trained as psychologists, could easily be confused with the physiologists, biophysicists, etc., at whose side they work.

RELATIONSHIPS BETWEEN VARIABLES

The psychological scientist seeks to discover the fundamental relationships between observable behavioral events and the conditions of the internal and external environment upon which they depend. These relationships, or laws, are generally stated in the form of functional equations, whether these equations are given in words or algebraically. The environmental conditions entering into psychology's laws are usually measured in units similar to those used by physicists, chemists, biologists, and other natural scientists, e.g., decibels (acoustic energy), minutes or hours (time), millimeters of mercury (pressure), etc., Similarly, the units in which behavioral events are expressed parallel those of the other sciences, e.g., amplitude of response (which may be stated in pressure, distance, duration, or electrical quantities, depending upon the particular response in question), number of responses per unit of time, latency, or delay of response, etc. Even when the psychologist cannot specify more than qualitative differences among the stimuli he presents and among the behavioral events he observes, he attempts to arrive at consistent and repeatable classification systems which, he hopes, will ultimately provide the basis for quantification.

DETERMINISM AND CAUSAL RELATIONSHIPS

The underlying presupposition of all scientists is that every event has an antecedent cause. Were this assumption not part of the scientist's basic philosophy he would be defeated even before he began his work. It does not even suffice for him to suppose that some things are determined by antecedent events while others may not be, since he would never be able to satisfy himself, or others, that a particular unexplained phenomenon was merely capricious rather than actually determined, but by a cause as yet unknown. While it is proper to speak of scientific determinism as an assumption, or presupposition, acknowledging by these words that it is logically impossible to prove that all events have antecedent causes (even events that have not yet been observed), it seems hardly fair to let the matter drop on so arbitrary a note. In actual fact, man's entire history attests to the probable correctness of the deterministic assumption. The number of phenomena that were, at one time or another, thought to be without lawful cause, but are now known to result from antecedent events, has grown astronomically and continues to grow even as these very words are read. Consider, for example, how recently Pasteur's work established the germ theory of disease. The scientist must really bend over backwards in his desire to be completely honest when he says that determinism is merely something that must be assumed rather than that the evidence overwhelmingly favors it.

Independent and Dependent Variables

In the climate of this deterministic assumption, the relationships that psychologists seek to discover between environmental conditions and behavioral events are ultimately causal, i.e., specifying a cause and an effect. While it is true that the mathematical use of functional relationships need not imply cause and effect, the scientist is not content, except temporarily, to note merely that two things tend to vary concomitantly. He wishes to go further and discover whether deliberate variation of one results in change in the other, or, when matters are not so simple as this, to discover what it is that he must vary to produce concomitant variations in his two original items of interest. When he can demonstrate that some condition of the internal or external environment stands in causal relation to some observable behavioral event, so that he may alter the former deliberately and observe concomitant alterations in the latter, he refers to the manipulated environmental condition as an *independent variable* and to the affected behavior as a *dependent variable.*

This use of the terms "independent variable" and "dependent variable" is somewhat narrower than that which is current in mathematics and, even, to some extent, in psychology. The mathematician is not concerned seriously (nor should he be) with the attributes of the real world for which his symbols may sometimes serve as substitutes. The psychologist who attaches the terms "independent variable" and "dependent variable" to conditions or events arbitrarily is frankly and explicitly disavowing an immediate interest in causality and would probably be better advised to dispense completely with both terms in such situations. For the present purpose it is desirable to restrict the application of the term "independent variable" to those manipulable environmental conditions that are either known (on the basis of prior evidence) or thought (i.e., hypothesized) to stand in causal relation to some observable behavioral events. Since a deterministic assumption asserts that every observable behavioral event is the effect of some antecedent cause or causes, we will hardly restrict ourselves when we apply the term "dependent variable" to those behavioral events under study in relation to particular independent variables, all other behavioral events being *potential* dependent variables.

Levels of Explanation

The unity and continuity of science, described above, gives rise also to a recognition that many different levels of scientific conceptualization are available and, in fact, many are employed by different scientists. In the field of psychology, for example, some scientists are satisfied to determine and describe relationships between the properties of environmental and behavioral events in the form of relatively simple, so-called low-level, functional relationships. Such an investigator might report, for example, that the main conclusion regarding an experiment was that some property of behavior (e.g., frequency of eye-movement) was found to vary reliably in relation to the frequency of occurrence of novel stimulus events. Another investigator, however, might wonder whether the relationship between eye-movement and novel stimuli was in any way capable of being understood in conjunction with other reliably observable concomitants of novel stimulation, such as cortical alpha rhythms, or in conjunction with other conditions known to influence eye-movement, such as degree of excitement.

In other words, some psychological scientists move back and forth quite easily between the directly observable empirical level and a level of conceptualization which may be one or more conceptual steps removed from actual observation. The latter investigators usually attempt to determine relationships among the low-level relationships of separate experiments. Their conceptualizations represent the beginning of scientific theories, which, as they gain wide empirical support and are further elaborated, may become even more abstract and less directly connected to concrete experimental observations.

The value of such enterprises depends upon the effectiveness with which they direct scientific investigators to problems whose study will add intelligibly to the organized data-base, and, of course, upon the accomplishment of a gradual development of understanding of natural processes. At each level of abstraction, empirical observations which have never previously been made may be implied by new conceptualizations. Experiments are then conducted to verify these implications, or to invalidate the conceptualization. Discrepancies among related experimental results may become less enigmatic. New experiments may be suggested whose results may transform previously antagonistic findings into logically related and mutually supporting ones. The danger that premature and/or bad conceptualizing may generate substantial waste of experimental facilities, time, and effort, is minimized by strict adherence to the rule that carefully collected data are always "right."

To the extent that the observer's eyes and ears reflect natural events accurately and honestly, we may be sure of at least one thing—nature is never wrong. Our theories, however we may have become enamored of them and however successful they may have been previously in dealing with nature's mysteries, are much more likely to be in error than are our observations. The highest achievement of the experimenter is to employ scientific techniques and controls in order to make nature's answers to his questions most obvious. When he has done this, it matters relatively little whether the answer he observes is the one he hoped or expected would be found, or the one that would have provided confirmation of his theory. However laborious and time-consuming may be the efforts that will be required to revise his erroneous theories, he will have taken a definitive step forward as a result of his (albeit negative) experiment.

This is not intended to imply that all that is required for erroneous theories to be dropped is one conflicting observation. One of the most obvious lessons to be learned from the history of science (psychology included) is that incorrect theories continue to be adhered to, in spite of considerable negative evidence, until they are replaced by new theories. Continuing to work within the framework of an imperfect theory until a better theory is proposed merely acknowledges that unrelated facts do not constitute scientific knowledge (and, of course, that the old theory was, itself, based upon facts).

Relational Knowledge

In fact, an important aspect of the observation that scientific knowledge is conceptual, even though it is about observable reality, concerns the way in which it is relational. The sensory experience resulting from being struck in the eye by a drop of rain is not, itself, knowledge. Nor is the isolated observation of a rat running, nor that of a child crying. These separate events participate in the production of knowledge only when the creative scientist interrelates them with other events. Actually, it is not even the object nor the event, but the abstracted dimension or property of the object or event being studied, which enters into relational scientific knowledge. How many raindrops fell? How fast did they fall? How long did the rainfall last? How many times did the rat run before quitting? How fast did he run? Even more abstractly, how consistent was the rat's speed from occasion to occasion?

The knowledge-building task of the scientist requires that he first determine which properties or dimensions of behavior to consider. This determination, of course, will be guided by his original inquisitiveness about the particular subject matter. But there is something of the "chicken and the egg" in this development, since the scientist must have some idea of which dimensions he will study in order to formulate intelligent questions to ask about them, but he cannot decide upon which properties he will observe until he has already considered some question about them. In any case, having decided which dimensions may be relevant initially, he must then determine the relationships between these dimensions and those of other events. He may find, for example, that the speed with which an animal runs is highly dependent upon how long it has been deprived of food, or that the number of hours a person sleeps depends upon whether or not he has access to a clock or can see the daylight. On the other hand, he may find that the particular property of behavior he has abstracted and measured does not appear to be related to any other factors he can think to observe, in which case the dimension will be of no use to him.

ANALYTIC METHOD

Although isolated observation does not itself comprise knowledge, it is often the beginning point in the process whose end product is knowledge. The ordinary raindrop falling into someone's eye, or an apple landing on someone's head, may provide the spark for an information-seeking enterprise which will end only when the raindrop or the apple is no longer an isolated event but is a manifestation of a fully understood natural process. It is difficult to speculate about what it is about one event or another that makes it become the object of scientific curiosity. Rare events are no more likely candidates for scientific inquiry than are everyday events. Curiosity is more likely to be stimulated when little is known about a particular event than when a great deal is already well established. Yet, this generalization cannot be pressed too far, since an unlikely occurrence within a well understood context may be equally adequate as a stimulus for investigative activity. And, within the confines of fully organized scientific experimentation, the more that is known about certain events, the more one finds it necessary to learn still more about them. The highest achievements in science are almost always the consequence of long series of experiments, frequently going on in several different laboratories at once. Sometimes the best experiments may be those whose results suggest more questions than answers.

Scientists approach the task of acquiring knowledge about natural events with a highly specialized, skeptical viewpoint. Centuries of confused fumbling have taught them that immediate impressions about the workings of nature are often misleading, that things frequently are not at all what they may have seemed at first. Consider, as an example, how convincing the impression is that the sun travels around the earth. Although the scientist's approach is somewhat like that of a naive child who takes something apart "to see what makes it tick," it is also like that of someone who is watching the performance of a magician and carefully attempting to discover how he does his tricks. Because he is trying to do more than merely catalogue all of nature's aspects, however, the scientist is much more than a careful, skeptical observer. Since his goal is that of comprehending the relationships among the properties of objects and events, he must always be alert to the possibility that he has overlooked important properties. In fact, he is constantly faced with the necessity of thinking of new dimensions, new ways of looking at nature.

There is little doubt that the work of the scientist involves as much inventiveness and creativity as that of any artist or poet.

New Dimensions

Suppose that a brief but careful observation were made of an airplane flying overhead and it was noticed that the plane appeared to be moving somewhat "sideways," as indicated by the dashed line in Figure 1.1.

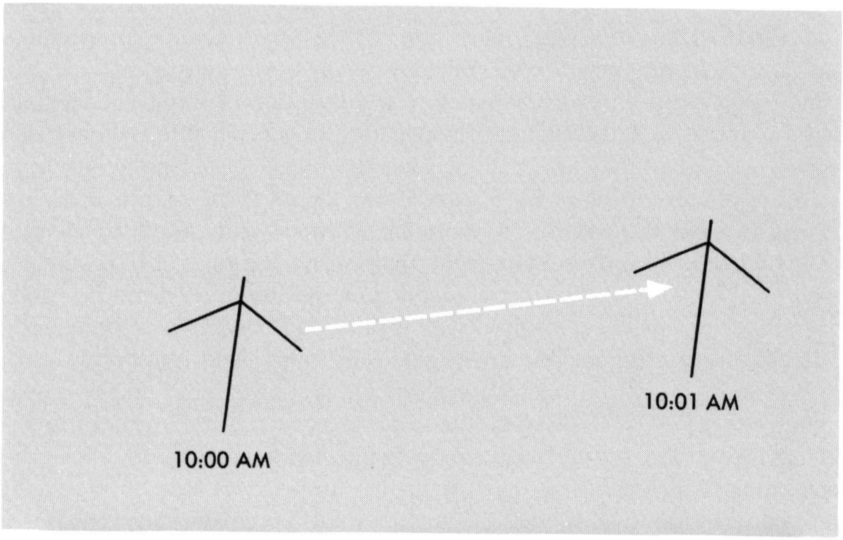

Figure 1.1 Pictorial representation of an airplane apearing to move sideways overhead.

With the naive knowledge that airplanes generally move in the direction in which they are "aimed," that is, in the direction indicated by their principal axis, the curious observer might be perplexed over what appears to be an incomprehensible event. By adopting an analytic approach to the phenomenon of the moving airplane, however, it may be possible to clarify this confusing impression. In its most simple appearance, the direction of movement of the airplane may be described by a single *vector*,[1] as is indicated by the diagonal solid line in Figure 1.2. By superimposing a set of orthogonal (i.e., perpendicular) coordinates upon the flat plane of the overhead scene, the single vector may be analyzed into two independent components of movement, each of which is sepa-

[1] A vector represents both the amount and direction of a force, with amount being represented by the length of the vector. In this example, the length of the vector represents the speed of the airplane in miles per hour.

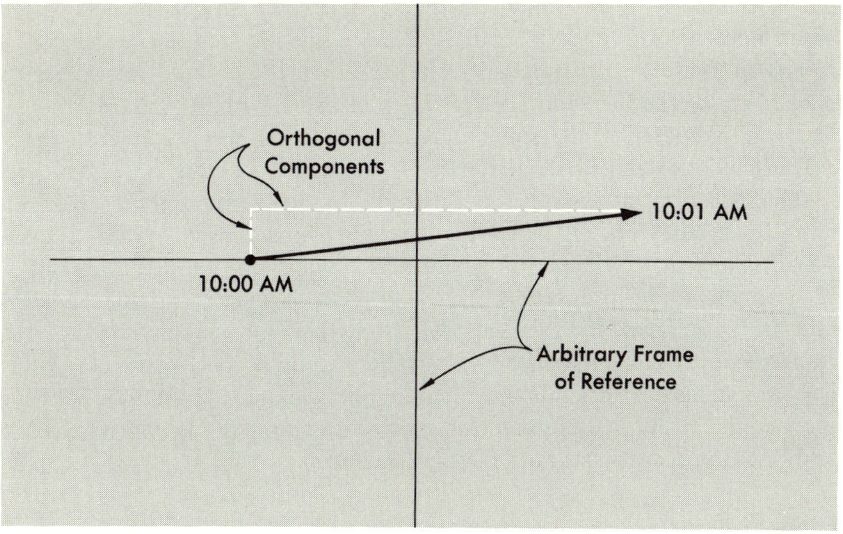

Figure 1.2 Analysis of naïve impression of airplane's movement into two orthogonal components of movement.

rately a vector, and the sum, or *resultant* of which is the original vector. It should be apparent that any line can be analyzed into two orthogonal vectorial components by the simple device of superimposing an arbitrary frame of reference. The orientation of the frame of reference is arbitrary also and can be adjusted to suit the observer's purposes. Only if the frame permits an analysis into meaningful components, however, is this type of analysis potentially useful.

One may well ask, then, why it is advantageous in the present instance to forsake the truism that the shortest distance between two points is a straight line and, instead, to describe the movement between two points as the resultant of two perpendicular components, traveling via the sides rather than the hypotenuse of the right triangle. In the case of the airplane it may be assumed that the plane's engines have the effect of pulling the plane in the direction it is aimed, but that some other force is acting on the plane to move it in some other direction at the same time. The ultimate path it takes, then, is the resultant of these two vectors. In order to fly the plane, in the present example, to the right and slightly upward, the pilot must aim it mostly upward. The wind, which is blowing from left to right, does the rest for him.

Note that it is easy to speculate about what the other force must be in the example of the airplane moving sideways, i.e., the wind, because the subject matter area is well-enough known to permit almost unequivocal resolution of the original problem. Suppose that wind direction and velocity were known and the "aim" and forward speed of the plane known, but that the plane's path could still not be described by means of these two components. What other force could be operating now to make the airplane behave inexplicably? Perhaps some mixed up kind of magnetism? Or some powerful invisible weapon whose effect is to deflect moving objects from their intended courses?

This example illustrates the logic by which explanations are built upon the framework of the particular dimensions chosen by the scientist for his analysis. In general, the task is always one of deciding which dimensions, which properties, are likely to be relevant, and then analyzing the superficial situation appropriately into its components. If the components fit into some meaningful scheme, and especially if this scheme has proven useful in analyzing related phenomena, the analysis may be successful. If not, a different frame of reference may be appropriate, or additional observational data may be needed.

Face Validity

Scientific analysis of the foregoing type is sometimes criticized on the grounds that the objects and events with which it deals do not seem to be made up of the peculiar dimensions scientists attribute to them. As was noted above, the sun does seem to be moving around the earth, rather than the reverse. It is for this very reason that objective analysis is employed by scientists, since things are so often not what they seem. In the field of psychology, this criticism has taken a special form, since complete scientific explanation of the behavior of man would contradict the primitive religious notion that man's behavior cannot be comprehended in terms of ordered relations but is, instead, ultimately capricious and freely modifiable voluntarily. Since this is at once a philosophical matter and an empirical question to be answered largely by data as yet uncollected, it will be sufficient for the present purposes to note that the order that man sees in the world around him is, after all, perceived by him. As he has become more and more able to observe his fellow men in as detached a way as he does the rest of the world around him, he has found order there too. There is no scientific reason to expect that this trend will be altered by future observations.

Humanistic Criticisms

Others have criticized the analytic approach of psychological scientists on the grounds that it tends to demean man and reduce him to hardly more than a very complicated computer–robot. This type of critic bases his argument on a highly elaborate, implicit preconception of what man is and what man is not. The scientist, on the other hand, begins his study of man with hardly any preconceptions as to his nature at all, other than the strong hunch that it is probably capable of being understood scientifically. A somewhat related criticism is that man will be turned into a machine by this approach to studying him, if he has not already been made one. Paradoxically, the historical roots of the scientific–analytic approach to the study of behavior are close to those of the current critics who complain that scientific psychology reduces man to a robot. Although one finds some of the elements of a mechanistic analysis of man's behavior in the writing of the Greek philosophers, there is little reason to doubt that the philosophical rebellion of Thomas Hobbes and John Locke, toward the end of the seventeenth century, marked the first clear beginning of the empiricist philosophy, from which scientific psychology springs. In a word, these British empiricists said that man becomes what he is through his experience, not because of his birth. The Aristotelian *tabula rasa,* the blank slate of man's infant mind, on which life's experiences draw the patterns of his ultimate character, was given a new significance by Locke. Because, if man is to become what his experiences make him, then the elucidation of the manner in which this developmental process takes place must have the highest priority. Thus, the beginning of a scientific analysis of man's behavior.

At the same time, if man is not born a beggar or a priest, a noble or a serf, then he is free to be whatever his experience may make him. Thus, at one stroke, British empiricism gave man his freedom, historically setting the stage for the beginnings of democracy and the present perversions of it which insist that man is so free that his behavior is capriciously unlawful, and created an analysis of his behavior that some claim will take this precious freedom away.

If there is a paradox in the possibility of a common origin of mechanistic scientific psychology and political democracy, there is still another paradox in the dogmatic insistence of some on man's a priori unknowability. To these critics of psychological science, it is an established characteristic of man that his characteristics can not be discovered by scientific analysis! They complain that scientific psychology may turn man into a robot; that it has already had some success in doing so. Their complaints are echoed by many different nonscientists. The college student who sees greater administrative concern for his not-to-be folded, spindled, or mutilated IBM card than for him is obviously in this group, as is the automobile driver who stops at stop signs only when other cars are visible. These critics are given to using words such as lockstep and automaton, as well as robot, when they give vent to their opposition to restrictions on their freedom.

Still deeper run the paradoxes. Consider the origin of the word "robot." One of Pavlov's contemporaries, the Czech writer, Karel Capek, coined the word for the title of his play, *RUR, Rossum's Universal Robots,* creating it from the Slavic word for worker. While it is true that these objects were manufactured for the machine-like purpose of performing all the distasteful duties required by civilization, including serving as soldiers in man's wars, it is equally true and of possibly greater importance that from their experiences and interactions arose the capacity for love, a most unrobotic trait. And Capek saw in his robots an even more "human" future than that of their extinct inventors. The "world was theirs" not only because they learned to love, but because their learned love was other-directed rather than narcissistic.

These introductory remarks are not intended to persuade the reader that scientific psychologists have correctly assumed that man is a machine, a lovable robot, and can be studied accordingly, but to suggest merely that the scientific analysis of man is not precluded a priori by nonscientific assertions regarding his unknowability or any other aspect of his nature. The scientist uses the methods of science because they have proved effective in the past in adding to his knowledge about natural events. Nothing that is now known about man supports the argument that he alone stands apart from the world in this respect. The humanistic anti-scientist, for reasons which will only be understood by means of painstaking scientific analysis, finds the scientific approach to the study of man objectionable. His behavior in this regard is simply another of the many manifestations of human complexity that make man so enticing an object for scientific study.

SUMMARY

Scientific knowledge is *public, objective, verifiable,* and *relational.* Starting with the assumption that every natural event has an antecedent cause, the psychological scientist seeks to discover causal relationships between environmental conditions and behavioral events, the former being referred to as *independent* variables and the latter as *dependent* variables. Different levels of conceptualization are employed by different scientists to bridge the gap between environmental and behavioral events, ranging from the most straightforward statement of actual observation to the most conjectural speculation about hypothetical intervening factors. The value of these theoretical speculations depends upon their effectiveness in guiding scientists to important new observations and in their contribution to his understanding of current observations. No matter how imaginative his speculations may seem, however, they must ultimately be capable of confirmation by means of *empirical* observation.

The scientist approaches his knowledge-building task in a highly analytical fashion, often looking beyond immediately obvious naive impressions to find possibly more fundamental relationships. He attempts to characterize each observed event in terms of its most relevant quantifiable dimensions, to discover the ways in which such dimensions may be related to one another. At times he may even invent dimensions which appear to enable simplifying relational statements to be made, sometimes seeming to represent the world in ways which are not immediately apparent. The earth doesn't seem to be moving around the sun, but it does. Also, an analytic approach to the study of man may create the impression that man is being treated as if he were a machine. As far as the scientist is concerned, however, no special assumptions regarding man's nature are made, other than that he may be studied and understood by the same scientific methods applicable to other objects and organisms in the world.

2

The Nature of Experiments

Perhaps the most significant difference between man and his subhuman relatives is to be found in his capacity and penchant for acquiring and transmitting knowledge of both his world and himself. From this point of view, it might be said of man that his history consists in substantial part of a progression of intellectual accomplishments which have taken him gradually out of ignorance and misbelief and into a broadening understanding of the workings of nature. Each rise to a new level of comprehension has been marked either by the shedding of some old restraint against acquiring new knowledge or the development of some new method for reorganizing old knowledge. That this has been and is still an accelerating process is testified to by the fact that, with each step out of ignorance, man's curiosity has grown rather than diminished.

EXPERIMENT AS A MEANS OF ANSWERING QUESTIONS

Viewed against the background of this quest for knowledge and understanding, the scientific experiment may well qualify as man's most noteworthy intellectual invention, since it represents the very finest technique he has devised to date for feeding his growing appetite for unambiguous answers to his questions about nature. Finally rid of rationalistic philosophy, increasingly impatient with uncontrolled naturalistic observation, and less and less content with mere probing and poking ("I wonder what would happen if . . . ?"), modern scientists have gradually developed and refined a method of inquiry that, at once, provides for an empirical method of evaluation and accords with the inductive–deductive logic in whose framework knowledge is organized. To remark that the basic principle of the experiment is too simple to justify so extravagant a claim, is, in the present view, approximately the equivalent of wondering why it took so long for the even simpler wheel to be invented.

Refined though it may be, the basic purpose of each and every scientific experiment is simply to answer a question about natural processes or events. It is distinguishable from other empirical information-seeking devices (e.g., looking out of the window to see whether it's raining) insofar as the question asked in experiments concerns a relationship between variables rather than a simple uncovering of factual reality, however sophisticated and expensive may be the apparatus used in the latter. Since, as was noted in Chapter 1, what the scientist typically means by knowledge is relational and transcends simple observation of isolated events, it should not be surprising that his experiments are styled to achieve this relational goal. This fundamental feature of experiments is present whether the relationships involved be between completely objective experimental manipulations (e.g., presence or absence of some drug) and uninterpreted behavioral measurements (e.g., rate of responding), on the one hand, or between operationally defined intervening variables (e.g., generalized drive) and hypothetical internal states (e.g., reaction potential), on the other.

Even while experiments differ in this way from simple empirical observations, however, their focus on sense-data for final testing is by no means a trivial characteristic. However gradually and even grudgingly, the scientist has come finally to the conclusion that only those of his ideas about nature that are both capable of empirical evaluation, and able to weather its cold scrutiny, are worthy of preservation. Since the knowledge we accumulate purports to be about the real world, it is not asking too much to require that it and all of its implications actually comport with empirical reality. This requirement is the raison d'etre of the experiment.

RELATIONSHIP BETWEEN EXPERIMENT AND THEORY

The foregoing remarks demonstrate that an extremely close relationship is assumed to exist between experiment and theory. This assumption, while seemingly in conflict with the fact that hundreds of perfectly good experiments have been done either in completely applied settings or by basic scientists who would deny ever having had a theoretical thought, is in substantial agreement with Boring's assertion, "Hypothetical constructs pervade science. They are the stuff of which it is made. The difference between being theoretical and empirical is mostly a question of how far the process of reification of the concepts has progressed"

(1953, p. 169). If Boring's term "reification" were changed to "explicit verbalization," the present position would be perfectly expressed. No experiments are or could be conducted in a complete knowledge vacuum, and theory is nothing more than an organized statement of current knowledge, the word "organized" referring to its explanatory as well as its constructive properties. Were no theory held regarding the manner in which the variables in a particular experiment operated, no controls could be devised, no basis would exist for deciding what to hold constant and what to ignore. It would be impossible to determine which aspect of behavior to observe without a background of organized knowledge and inference from it, nor which quantitative properties of the selected behavior actually to record and analyze. Even an experiment with a completely applied goal (e.g., Which formula should we use for next season's ice cream?) is effective only to the extent that it is planned in relation to what is already known, with varying degrees of certainty, about the relationships between several independent variables and the behavior of current interest (i.e., between the various characteristics of ice cream ingredients and variations in human taste preferences).

TWO TYPES OF EXPERIMENTS AND A QUASI-TYPE

A psychological experiment is a controlled set of observations on an objective measure of behavior (the dependent variable) under deliberately differing conditions (the independent variable). It seeks to determine either *whether* or *in what fashion* the independent and dependent variables are related, being particularly directed at elucidating causal influences of the manipulated independent variations on the observed changes in the dependent variables. In a manner of speaking, the independent variable involves what the experimenter does and the dependent variable involves what the subject does.

The first type of experiment mentioned above (*Whether?*) is known as a hypothesis-testing experiment. Its main question is answered when it has been determined that the independent variable does or does not influence the dependent variable. In psychological research this is by far the more common of the two main types. The second type of experiment (*In what fashion?*) is known as an estimation, or parametric, experiment. It seeks to describe the relationship between the independent and dependent variables in an exact quantitative form, hopefully as a functional equation. Estimation experiments are relatively infrequent in psychology, but it may be expected that they will become more common as the accumulated level of knowledge increases and the exact quantitative features of already identified relationships are sought.

In the area of problem solving, for example, the psychological scientist might ask whether the size of a group of individuals who are working together on a problem influences their effectiveness in arriving at its solution. For practical purposes, he may decide to use the time taken to reach the solution as the dependent variable in his study. To answer the simple *whether* question, it might only be necessary to employ groups of two different sizes, for example, groups containing two individuals, the smallest size group possible, and groups containing four individuals. The actual experimental plan might call for several different groups of the smaller size and several different ones of the larger size, with each group given one or more problems to solve. Or, it might be decided to use only one group of each size, with each group given many problems to solve. In either case, the results of such an experiment might lead the experimenter to conclude that group problem solving performance is related to group size (e.g., larger groups solve problems faster than smaller groups). This type of experiment is exemplified by a study by Taylor and Faust (1952), in which the popular game called "Twenty Questions" was employed by the subjects.

Estimation Experiments

Beyond answering the *whether* question, it may be important for the experimenter to determine in greater quantitative detail exactly how the variables of group size and problem solving performance are related. This changes the question being asked from the *whether* form to the *in what fashion* form. Studies on group problem solving performance, seeking to answer the *in what fashion* question, would employ groups varying in size over a wide range of values, for example, from two to five or more. These studies might show that improvement in group performance as group size increases reaches a maximum, beyond which further increases in group size fail to enhance performance and even impair it.

Quasi-Experiments

A third type of experiment, or *quasi-experiment,* is often conducted in psychology, in which an experimenter is interested in the relationship between a nonmanipulable independent variable and some observable property of behavior, but is not satisfied with mere naturalistic observation. In this category is a wide variety of studies in which variations in the independent variable already exist naturally and its levels can be measured or, if the variable is qualitative rather than quantitative, its categories reliably distinguished. The experimenter measures or identifies the differences in the independent variable, most often some property of individual subjects, and selects groups of subjects that are representative of either certain arbitrary levels or categories of the variable or are representative of the natural pattern of its variation. Observations are then made of the dependent variable in the selected groups, under controlled conditions.

A typical example of this type of quasi-experiment was done by Taylor (1951), in which a large group of subjects was given a paper-and-pencil test of "manifest anxiety." From these subjects two groups were selected containing individuals with extremely high or low scores. These individuals were then run in an eyelid conditioning procedure. The results of the conditioning part of this study indicated that the "high anxious" subjects conditioned more readily than the "low anxious" subjects, which was interpreted by Taylor to mean that the drive level associated with anxiety tends to facilitate conditioning.

However, without manipulative control of the independent variable it cannot be ascertained that the relationship revealed in the experiment is either causal or direct. For this reason, it is inappropriate to speak of the influence of manifest anxiety on classical conditioning, where the former is inferred from responses on a psychometric test and the latter inferred from proportions of conditioned responses. For the same reason, it would be inappropriate to speak of the effect of classical conditioning on manifest anxiety had subjects been put into groups on the basis of percentages of conditioned responses and then their scores on an anxiety test obtained. As Spence (1956) has pointed out regarding "experiments" of this type (labeled R–R, or correlational), both variables are response-inferred and both could easily be dependent upon one or more other, unknown independent variables. The most that can be said about their relationship is that it is either direct or indirect. It is, at least, reliably observable under more or less controlled conditions.

Accordingly, quasi-experiments involving selected levels or categories of the independent variable should be distinguished from the far more primitive procedure of ordinary naturalistic observation, since the more controlled selection method of quasi-experiments is far more likely to yield unambiguous information regarding the relationship in question, whether or not it is causal. In both manipulation and selection experiments, the conditions under which the dependent variable is observed are artificially regulated and the possible influence of other relevant variables may be controlled or concomitantly studied. For this reason, and because the major design principles to be presented in the following chapters are relevant to both, no distinction will be made between experiments involving independent variables defined by manipulation and quasi-experiments involving independent variables with selected levels.

As has been indicated in the foregoing, experiments flow naturally from the deductive elaboration of explanatory theories, providing both the factual substrate from which the theories are inductively built and the eventual testing ground of their soundness. On the one hand, inductive generalization from prior experiments is employed to create a theoretical explanation and, on the other, deductive generalization from a theory establishes the means by which its adequacy is measured. Without the latter, the theory stands weakly as an ex post facto fabrication; thus the emphasis on empirical testability in formal analyses of theory construction. No matter that the inductive generalizations assembled in the theory came directly from observed data; unless the concepts that bind them together have empirically testable implications (whether or not they have yet been tested), the theory is potentially worthless scientifically. Without the former, explicit in the form of postulates and theorems or implicit but functioning as vague focusing agents, experimentation would be a senseless, aimless exercise; or experimentation simply would not be.

The detailed manner in which the psychological experimenter goes about the task of analyzing his theoretical notions so that operationally specified independent and dependent variables emerge for experimental use will be considered in Chapters 3 and 4. For the present we will assume that these decisions have more or less been made, in order to consider beforehand some of the logical properties of experiments.

HYPOTHESIS-TESTING EXPERIMENTS

Experimental Question

The hypothesis-testing experiment begins with the question, "Is there a relationship between some independent variable (X) and a dependent variable (Y)?" Typically, and desirably, experimental questions are put in very general form. The purpose is to begin at a level of discourse that clearly transcends the specifics of any individual experiment. No effort should be made to provide operational definitions in initial statements of the experimental question, although experimenters frequently find themselves unable to avoid a minimum degree of narrowness. In the present view, as will be clarified below, a nonspecific beginning is to be preferred.

A few examples of explicit statements of experimental questions, selected at random from a recent single issue of the *Journal of Experimental Psychology,* will give some indication of their range of specificity–generality. "The present experiment was an attempt to clarify the role of the number of choices in maze learning." "The purpose of the present experiment was to determine the effects on retention of systematically varying the amount of uncertainty associated with the decoding of trigrams." "The following questions were of interest: Does discrimination training produce a displacement of the mode of the generalization gradient from S+ in the direction away from S−; is the displacement of the mode inversely related to the S+, S− difference; and does discrimination training increase the steepness of the generalization gradient?"

In each of these examples some significant terms are used without even the slightest pretense of operational definition, while, in some of them, key terms are stated almost precisely enough to enable us to guess the details of the experimental procedure. Since the experimenter usually wishes to ensure that his experiment will retain the broad implications that it possessed when he first formulated it as a question, some pains may have to be taken to avoid too rapid a reduction to specific operational specifications.

Experimental Hypothesis

Experimental questions of the more or less general type discussed above, while playing a role in the experimenter's developing conceptualization, frequently are omitted entirely from experimental reports in scientific journals, if not even from the experimenter's own explicit thoughts. This is so because experimental questions are often formulated even initially as hypotheses. Even when given explicit interrogatory form, as in the last of the examples quoted above, the question may actually be a prediction of outcome, frequently specifying both magnitudinal and directional details. The experimental hypothesis is nothing more than a re-verbalization in declarative, and sometimes, directional, language of the relational substance of the experimental question.

The advantage of going through these two somewhat redundant stages, (1. Question, 2. Hypothesis) is found in the fact that the experimental hypothesis can be couched in language of a much more operational type than the experimental question should contain: Thus, the hypothesis serves a transitional function from the question to the procedures, and enables both the experimenter and his potential audience to retain the connection between his general theoretical ideas and the concrete operations by means of which he proposes to test them. In addition, directional statements of experimental hypotheses, especially when the explanatory theories from which they emerge are highly developed quantitatively, represent the forerunners of estimation experiments, if not achieving that level themselves.

Null Hypothesis

The experienced experimenter is usually already aware at the time the experimental hypothesis is first stated of exactly what kinds of differences between measures of behavior are to be examined in attempting to confirm or invalidate the hypothesis. Depending upon the degree of experimental control attainable in the behavioral domain under study and, to some extent, upon the scientific philosophy and values of the experimenter, methods of statistical inference may be required as adjunctive support for any conclusions regarding the effect of the independent variable. Since the role of statistics and statistical logic in psychological experimentation is treated in detail in Chapters 6 and 7, only a few introductory comments regarding decisions to reject or not to reject the null hypothesis will be made here.

When statistical inferential methods are used, the experimenter is at pains to justify in his description of procedure and his presentation of results that their logic reasonably applies to his situation. The most common approach to this problem is to start with the assumption that the independent variable has no effect whatever on the observed behavior, or, more broadly, that no systematic factor has influenced the behavior. This assumption, when elaborated logically into a model of the manner in which unsystematic variations in the dependent measures could occur, is known as the null hypothesis.

If the experimenter has employed the principle of randomization, discussed in Chapter 6, in such a way that only chance could influence the dependent variable under the various experimental conditions (and, given the assumption that the independent variable has no effect), the question may then be asked, "What is the likelihood that the difference in behavior observed (usually a difference between sample means or other measures of central tendency) could have happened solely as a result of chance?" Obviously, what is needed is some way of estimating how variable the dependent measure is when nothing systematic is influencing it. If the experimenter can generate information of this type, he may then be able to specify with some accuracy how often differences of various amounts might be expected by chance between samples of the size he used. It is then only a fairly simple step to a specific statement regarding how often his observed difference could be expected by chance. An accurate estimate of the chance probability of obtaining a difference at least as large as his observed difference is all that statistical inference can give him. The actually observed difference remains what it was before the statistical manipulations were undertaken. But, armed with an accurate estimate of the likelihood that his results could have occurred by chance, the experimenter is in a position to argue with far greater confidence about the relationship studied in his experiment than he would be without it.

Straightforward and simple as the foregoing series of statements may appear, every student of psychology is well aware of the vast complexity of statistical technology that lurks in its unsaid background. That the actual implementation of the logic of statistical inference is a more difficult proposition than is that of comprehending it may be due partly to the fact that the chance fluctuations in the dependent variable that must be estimated may arise from several sources. Among the more obvious of these sources are the unsystematic errors made by the experimenter or his machines as the behavior is recorded, the unsystematic fluctuations in uncontrolled and unidentified variables that may influence the dependent variable, as well as possible unsystematic oscillations in the actual state or process under study. Furthermore, the extent of variation stemming from chance factors depends upon whether randomly assigned, different individuals are studied under the various levels or categories of the independent variable, whether deliberately matched individuals are used, or whether the same individuals are studied under all conditions.

Rejecting the Null Hypothesis

The experimenter does not assume the null hypothesis because it has credibility for him. He does not actually believe it to be a reasonable possibility. Under the conditions of most hypothesis-testing experiments, the null hypothesis is a generally unlikely state of affairs. Experienced researchers do not conduct elaborate experiments involving independent variables whose effects are actually expected to be nonexistent, except on the rare occasion when someone else thinks an effect may exist but the experimenter does not. The null hypothesis is assumed solely for the purpose of devising a test whose goal is to reject it. Only by starting with the assumption that the independent variable has no effect at all, can the experimenter construct a model of chance outcomes in the framework of which the actual outcome may be comparatively evaluated.

Inability To Reject the Null Hypothesis

Once it is understood fully that the null hypothesis is the experimenter's own invention, set up for his convenience in order to permit statistical evaluation of his results, it should be clear that he never concludes an experiment by accepting the null hypothesis. Since there is considerable disagreement and confusion associated with this point, it will be worthwhile to examine it in detail. Suppose that the statistical test employed reveals that under the assumption of the null hypothesis a difference as large as that observed could have occurred by chance fewer than one time in one hundred. This probability statement, referred to as *alpha,* or the significance level, is based on the model of chance outcomes thought by the experimenter to be an appropriate reflection of the unsystematic influences operating in his experiment. When an experimenter reports that his results have achieved, say, the 0.01 level of significance, he is saying simply that he estimates that the chances are only one in one hundred or fewer that the null hypothesis could actually be true. If the model on the basis of which this statement is made is really appropriate for his experiment, i.e., if the null hypothesis were really true, the results could only have occurred that infrequently. Since there exists a more likely and reasonable alternative to the null hypothesis, namely, that the independent variable was responsible for the observed difference, the experimenter rejects the null hypothesis in favor of that alternative. He concludes that a relationship between his independent and dependent variables exists. Depending upon the exact wording of his experimental hypothesis, a statistically significant difference may be taken as confirmation or invalidation of it.

Suppose, however, that the statistical test used by the experimenter indicates that a difference as large as was observed could have occurred by chance as often as twenty times in one hundred, assuming the null hypothesis. Very few experimenters would be willing to reject the null hypothesis in favor of some experimental alternative under these circumstances. The possibility that chance alone was responsible for the outcome is too great to be discounted. In such cases the experimenter may be disappointed or frustrated, but he does not conclude that the null hypothesis was true. He does not accept the null hypothesis; he simply is unable to reject it. His statistical test has told him that his results could have happened by chance 20% of the time, not that they did happen by chance.

At this point the experimenter must be aware of how and why the null hypothesis was invented in the first place. Failure to reject the null hypothesis does not mean that there are not many hypotheses with far greater likelihood of being true. The most likely ex post facto hypothesis in the situation just described, based on the data obtained, is that the independent variable had some effect. The experimenter is unable to reach this conclusion only because the likelihood is too great that chance was responsible for him to risk it, not because the likelihood that chance was responsible is greater than the likelihood that the independent variable was responsible.

"The most likely hypothesis in the situation just described . . . is that the independent variable had some effect." One way to elucidate this statement is to consider how one would bet on the outcome of a complete repetition of the experiment. Anyone who seriously argues that failure to reject the null hypothesis implies acceptance of it (or "retention," a more popular term among statisticians) can be asked to place his bet on the predicted outcome of no difference, in opposition to a bet on a repetition of exactly the difference that was obtained the first time. Whichever is closer to the actual second outcome wins the bet. No statistician would make such a bet, since it is well-known that the best prediction of the obtained difference in the second study is exactly the difference that was obtained in the first study; both smaller and larger differences are less likely second outcomes. Almost no one would accept the null hypothesis as the most accurate prediction of what would happen were the experiment repeated, and one can hardly fail to recognize that this is equal to accepting the null hypothesis as the most accurate description of what actually did happen.

TYPE 1 AND TYPE 2 ERRORS

If the null hypothesis is rejected by the experimenter, the possibility always exists that he has made a mistake and that the null hypothesis is really true. Mistakenly rejecting a true null hypothesis is known as a Type 1 error, and the probability that this has occurred is given by *alpha*, the significance level actually obtained. That is, when the experimenter states that his results are significant at the 0.01 level, he is reporting how often these results could be expected to happen by chance, in this case very infrequently. Were the null hypothesis actually true, and were the experiment exactly repeated one hundred times, a difference at least of the size in question would occur approximately once. In psychological research, results that achieve the 0.05 level are often referred to as significant and those that achieve the 0.01 level as very significant. These standards are obviously arbitrary rather than being hard and fast rules.

If the null hypothesis is actually false but, for some reason, the results do not achieve a level of significance sufficient for its rejection, a Type 2 error has occurred. It should be obvious that Type 2 errors would never occur if the experimenter were willing to report all observed differences as significant, i.e., if the experimenter were not attempting to minimize the chances of Type 1 errors. In this sense the two types of errors are reciprocally related, since the probability of a Type 2 error increases as lower and lower probabilities of a Type 1 error are required by the experimenter for rejection of the null hypothesis. Both types of error are clearly undesirable, but which is the more dangerous in a particular situation depends upon the consequences of each type of error in that unique situation. In actual experimental practice, only the probability of the Type 1 error is actually known.

A Priori Choice of Alpha

Experimenters are frequently urged to make a decision regarding the significance level to be used in an experiment prior to the collection of any data and to adhere to that decision thereafter. This injunction stems largely from the fact that experimenters are only human, and they may be willing to alter their standards regarding rejection of the null hypothesis in a lenient direction after having invested all of the time and effort involved in doing the experiment. In the main, this requirement seems to have been overemphasized, since the 0.05 and 0.01 standards are quite widespread in the field and often are employed by editors of scientific journals among the criteria for acceptance or rejection of research reports for publication. In addition, the mistaken impression sometimes exists that only by committing to writing beforehand his chosen level of significance for a particular experiment, can the experimenter actually report it accurately. This dictum frequently includes the corollary recommendation that the results should only have attributed to them the significance level that was chosen before the experiment was conducted, even when the probability actually obtained turns out to be considerably smaller than even the conventional standards.

If the assumptions underlying his statistical procedures were adequately justified, nothing that the experimenter writes down beforehand can alter accuracy of the obtained estimate of significance. Having the significance level written down beforehand is not one of the assumptions required. Furthermore, failure to report the actually obtained significance level when it exceeds the minimum required by the experimenter or by convention is a form of suppression of data that conflicts with the accepted standards of empirical science. How can the reader of a research report be expected to arrive at an informed assessment of its results if he is not told what the chances are that they could have occurred under a true null hypothesis? The observed data, the statistical tests made on them, and the probabilities associated with these tests are all part of what was found in the experiment and should be reported as such.

SENSITIVITY AND POWER

The goal of an experiment is to detect the presence of relationships when they actually exist. Because of the presence of error, a necessary additional goal is to provide a means for estimation of its extent. The smaller the differences in the dependent variable measures associated with the independent variable, the more sensitive the experiment must be in order to detect them. Highly sensitive experiments permit the detection of very small differences, while experiments with low sensitivity can only detect very large differences.

Since there is a tendency to confuse the term "sensitivity" with the related term "power," it is useful to make the following distinction between them. Sensitivity is a property of experiments, while power is a property of statistical tests, although both are usually defined in terms of either detection of differences or rejection of null hypotheses. The confusion arises in part because one of the ways in which sensitivity can be increased is to employ a more powerful statistical test. The reverse, of course, is logically meaningless. All other things equal, the more powerful the statistical test used in an experiment, the more sensitive the experiment. It goes without saying that the experimenter should employ the most powerful statistical test available to him within the limits established by the assumptions he can justifiably make.

Mathematical statisticians typically describe the properties of statistical tests in the form of a *power function.* A power function is a representation of the relationship between the probability of detecting a difference and the magnitude of the difference. The operating characteristic of a power function, a reflection of the way in which the probability of detection of a difference changes as the size of the difference changes, is the main basis for asserting that one statistical test is more powerful than another, against equivalent null hypotheses. Thus, certain statistical tests, known as *maximum likelihood ratios,* are generally more powerful than others, the latter usually involving assumptions that are easier to justify in the experimental context. The most common tests used by psychological experimenters, the t and F tests, also happen to be the most powerful in situations in which they may be used appropriately.

Increasing Sample Size

Besides using the most powerful statistical test whose assumptions can be met, the experimenter may be able to increase the sensitivity of an experiment in more direct ways. One way to increase sensitivity, of course, would be to increase the size of the sample. Obviously, an experiment would be sensitive enough to detect even the smallest difference if the entire population of measurements were available for consideration, since, in this case, its results would merely be a description of the entire population.

Because of obvious economic factors (and because "population" is often not really a finite number of possible observations, but is infinite and frequent hypothetical), the aim of the experimenter is to achieve a level of sensitivity sufficient to permit detection of real relationships, meeting conventional significance standards, with as small a sample as possible. This point is often lost or even misrepresented in the midst of arguments regarding the beneficial effects on sensitivity of increasing sample size. Thus, while it is surely true and important that sensitivity increases as a positive function of the square root of sample size, skillful planning of experiments does not consist in merely arranging to use large samples.

REDUCING ERROR DIRECTLY

The unsystematic sources of variation in the dependent measures, which tend to mask the effects of independent variables, are generally unknown in any particular experiment. But they are unknown only in the most narrow sense that they cannot be separately identified and their influence measured or regulated individually. In another sense, the main sources of error are available for their direct reduction. For example, unsystematic fluctuations in attention or motivation would seriously inflate the error variance in many experiments involving human subjects were it not possible to give them precise instructions and to reward them with money or class points for their participation and cooperation.

Controlled experimental environments, such as soundproof chambers, also serve to eliminate unsystematically variable sources of stimulation in experiments in which controlled stimulation is critical. In psychophysical experiments, subjects may have to be very highly trained to enable them to ignore irrelevant and usually unsystematically variable influences while they are responding to the stimuli of experimental interest. And, in operant conditioning studies, considerable time and effort are expended to ensure that the effects of independent variables are studied against a background of stable behavioral baselines.

Individual Differences

By far the most serious and, yet, the most well-known source of unsystematic error in psychological research is the fact that the individuals who serve as subjects may differ substantially from one another and from moment to moment in the very response characteristics under study, frequently even before the experiment is started. These differences may be of sufficient magnitude in certain research areas as to exceed even the largest differences likely to be produced by the independent variables studied, e.g., in studies of factors that may influence attitudes toward members of ethnic minority groups. It is not accurate to say of such situations that the effects are thus of little interest. Even when the range of individual differences in the dependent variable is not extremely large, it still may be a significant experimental problem because of the inaccessibility of anything but very small samples of subjects for experimentation.

The use of matched pairs of subjects, littermates, or repeated measurements of the same subjects in several experimental conditions, each of which is an application of the blocking principle, covered in Chapter 9, all are devices intended to reduce error associated with individual differences in the dependent variable. In addition to these direct, experimental methods for reducing error from individual differences, *indirect,* statistical, techniques of adjustment of the dependent variable are sometimes used. The most common of these will be discussed in Chapter 7.

COMPARISONS OF THEORETICAL AND OBTAINED CURVES

Recently, several experiments have appeared in the literature, in which predictions of a specific, quantitative sort have been generated from mathematical models of particular psychological processes, e.g., predictions of asymptotic proportions of dichotomous guessing behavior under various manipulated levels of probability of occurrence of each outcome, with the view in mind of comparing actually obtained with theoretically expected functions. As Grant (1962) has noted, in his discussion of this question, a tendency is discernible on the part of some of the experimenters who have done this work to aim their statistical tests toward the goal of accepting rather than rejecting the null hypothesis. They are typically interested in demonstrating that the obtained data-points do not deviate sufficiently from those predicted by their theories to warrant rejection of the null hypothesis and invalidation of the theory upon which it was based.

One of the best ways to ensure that rejection of such null hypotheses will be highly unlikely is to conduct the experiment on a very small sample in a deliberately uncontrolled way, so that the estimate of error will be extremely large. The more insensitive the experiment, the less likely the rejection of the theory, a rather paradoxical state of affairs! One way to modify such experimental tests of theoretical predictions, which preserves the possibility of using rejection, rather than acceptance of the null hypothesis, is to redirect the original statement of the null hypothesis so that it focuses on agreement rather than disagreement between prediction and the outcome of an experiment. A test of the hypothesis of zero correlation between the predicted and obtained data-points meets this requirement. Rejection of the null hypothesis of zero correlation can result in acceptance of the alternative hypothesis that the obtained and predicted points are significantly related. This approach puts the responsibility on the experimenter to conduct a sufficiently sensitive experiment, which is as it should be. As Grant argues, estimation procedures are ultimately those that will be preferred in tests of quantitative theories, since they are flexible enough to identify both the degree of agreement between theory and data and the most important points of disagreement.

ESTIMATION EXPERIMENTS

The procedure known as setting confidence limits is basic to experiments in which interest centers more on obtaining an accurate estimate of the true value of the dependent variable at each selected level of the independent variable than upon merely demonstrating that the dependent variable differs as a function of the independent variable. Basically, this entails determination of the standard error of the mean (i.e., the standard deviation of sample means), followed by the marking off of an interval (e.g., plus or minus three standard errors on either side of the sample mean), within which the chances are known that the population value must lie. In an experiment involving two arbitrarily selected or manipulated levels of an independent variable, a confidence interval would be set up around the observed mean difference, providing a range within which the true difference is likely to be found. If the confidence limits are located three standard errors on either side of the obtained mean difference, the chances are fewer than one in one hundred that the population value lies outside of them. The smaller this interval is, the more precise the estimate of the parameter, since precision is inversely related to the size of the standard error.

As was noted previously, estimation experiments have not been conducted very often in psychology, probably because the state of knowledge has not been advanced enough to justify them (cf. Meehl, 1967). There is some reason to fear that cultural lag may also be responsible, since the hypothesis-testing tradition is very strong at present and it may even unfortunately be the case that parametric studies are looked down upon as less challenging and less demanding of experimental creativity than those that test hypotheses. It is to be hoped that prejudice of this type will not persist long enough to impede psychology's progress as a quantitative empirical science.

RELIABILITY OF MEASUREMENT AND OPERATIONISM

Over and above the innumerable chance situational–experimental factors that may influence the dependent variable, some of which were mentioned above, errors inherent in the measurement process itself may be a major source of error. This type of unsystematic error variation is traditionally referred to as *unreliability,* the term coming from the field of psychological testing. Although the term is often applied to all unsystematic variation, regardless of whether it is or is not inherent in the measurement process (for example, differences between sample means which are too small to justify rejection of the null hypothesis are frequently called "unreliable" differences), it may be reserved exclusively for the measurement process, particularly for errors associated with the measurement instrument, when one can be identified explicitly.

An example of unreliability of measurement of the type in question would be the fluctuations in recordings of response times resulting from the mechanical slippage in the clutch of electric timers. Repeated clockings of exactly the same time periods, e.g., produced by a highly accurate electronic device, can be observed to vary, solely because of the fact that the electromechanical unit that stops the moving hand of the timer has a variable tendency to slip. Sometimes it slips very slightly; but sometimes it slips considerably. In this case, all of the errors are in the direction of overestimation and the shape of their distribution is asymmetrical.

A more obvious example would be the unsystematic tendency of the subject sometimes to mark unintended response choices in a paper-and-pencil test. The intrusion of random electrical signals into the recordings of bio-electric data, inflating some measured responses and attenuating others, is another example.

Operational Definitions

Measurement has been defined as the assigning of numerals to objects or events according to a set of rules. A precise statement of the rules provides what is known as an *operational definition* of what it is that is being measured. A major potential source of unreliability in the dependent variable is associated with a lack of objectivity (explicitly or implicitly) in the rules of measurement.

In an eyelid conditioning experiment, for example, the dependent variable to be studied may be the percentage of conditioned responses made by the subject in a certain block of trials. The question immediately arises as to what is scored as a conditioned response and what is not. Because of the way in which movements of the eyelid are transmitted to the permanent record, deflections may occur on the response record that are not due to actual lid closures but to other sources, e.g., movement of the forehead. For reasons of this type, it is conventional to require some minimum amount of movement, e.g., one millimeter, before a response is counted. The minimum must not be so high as to eliminate more than just a few actual responses, otherwise systematic bias may be introduced into the data. But, some minimum will tend to eliminate isolated random fluctuations and increase the reliability of the percentage measure. Similarly, the time period within which responses are counted may be narrowed down to reduce intrusions of nonconditioned responses into the data. Since response deflections other than those assumed to be conditioned responses are likely to be influenced by variables that may be uncontrolled, temporal restrictions in the definition of conditioned responses are likely to increase reliability of measurement. Again, however, it must be ascertained that a minimum of conditioned responses are eliminated.

Operational Definitions and Repeatability

A significant reason for urging operational specification of the dependent variable in terms of measurement rules is that procedural *repeatability* of the experiment is enhanced greatly thereby. To the extent that the experimenter has provided a set of concrete steps to be taken in measuring the dependent variable, it is possible for others to repeat his work and, perhaps more importantly, to follow-up on it. Whether his dependent variable has any psychological meaning beyond that given by its manner of measurement, a point which is disputed by advocates of "operationism," concrete measurement rules at least make it possible to identify its operational meaning.

Explicit statements regarding the reliability of the dependent measures are seldom given in experimental reports, other than those implicit in the statistical tests described. This is only partly due to the fact that many experiments involve only a single measurement per subject, which precludes some of the more common reliability estimates derived from psychological test theory. Even in such cases, however, it might be useful to develop appropriate reliability coefficients that would indicate the proportion of variance in the dependent measures due to nonrandom factors, such as an intraclass reliability estimate (Haggard, 1958). In cases involving repeated measurement of the same subjects, of course, either an intraclass measure or a measure of the test-retest type could be reported, but, again, seldom is.

Reliability and Interpretation of Results

The reason an explicit reliability estimate would be a useful adjunct to experimental reports and to experimenters during the conduct and planning of experiments, is that failure to reject null hypotheses may sometimes more readily be attributed to the unreliability of the dependent variable rather than to the ineffectiveness of the independent variable. If an experimenter has devised some measure of behavior that turns out to have low reliability and has estimated that reliability, more time might be spent on the problem of increasing the reliability of the dependent variable than is typical. As will be suggested in greater detail in Chapter 4, this problem is considerably more acute in experiments involving complex, derived measures of behavior, or ratings of behavior made by persons, than it is in experiments in which physically defined dependent measures, e.g., time, force, etc., are used.

Experimenters frequently attend primarily to testing hypotheses regarding differences in behavior and, accordingly, are quite satisfied with the reliability of any variable in which significant differences are found. An additional reason for the lack of emphasis on estimates of the reliability of the dependent variable is the tendency for experimenters to be disdainful regarding the use of correlational logic and methods in their work (Cronbach, 1957). This is unfortunate because the concept of experimental control, usually reserved for reference to the identification and restriction of influence of factors that affect the dependent variable, takes on broader meaning when it is coupled with that of reliability. Estimates of reliability should be direct reflections of the extent to which control has been effective and their use would take experimental psychologists beyond their present informal stage regarding this issue.

PROBLEM ANALYSIS

For the purpose of the foregoing presentation, it was assumed that the experimenter had already decided exactly which independent variables to manipulate or select and exactly how to measure the dependent variable of interest. Obviously, a major part of his work is involved in arriving at these decisions before getting down to the specifics of running the experiment. The term "problem analysis" [1] will be used to refer to the method by means of which the experimenter's theoretical ideas are assembled and reassembled into the form of an experimental question and an eventually more specific experimental hypothesis. At its broadest level, the logic of problem analysis is largely analytical and deductive, insofar as the choice of the independent and dependent variables is concerned. This topic will be discussed in Chapters 3, 4, and 5. In Chapters 3 and 4, in addition, the mechanics of operational specification of dependent variables will be considered; Chapter 5 is reserved for the problem of choosing the independent variable.

SUMMARY

The scientific experiment is man's most effective technique for answering questions about natural processes or events, seeking to determine either *whether* or *in what fashion* variables are related. *Hypothesis-test-*

[1] This term is adopted with modifications from the lectures of W. W. Grings.

ing experiments attempt to discover whether or not some independent variable influences some dependent variable. *Estimation,* or *parametric experiments,* produce descriptions of the detailed quantitative manner in which variables are related, after the existence of a relationship has been determined. *Quasi-experiments* are similar to hypothesis-testing experiments, except that they involve independent variables whose levels are selected from already existing variation rather than being deliberately manipulated by the experimenter. For this reason, it is inappropriate to speak of causal relationships in connection with quasi-experiments.

The hypothesis-testing experiment begins with a general query, called the *experimental question,* regarding the possible relationship between the independent and dependent variables. When the variables suggested in the experimental question are given *operational definition* and the question is reverbalized in declarative rather than interrogatory form, the *experimental hypothesis* is produced. The experimental hypothesis is an unambiguous assertion regarding the expected results of the study, even to the point of specifying their anticipated quantitative and directional details. To evaluate the possibility that the obtained results may have been due to chance rather than his experimental manipulations, the experimenter adopts what is called the *null hypothesis,* that is, he assumes that chance alone was responsible for the results. This null hypothesis is elaborated into a statistical model of chance outcomes, which permits the experimenter to state the likelihood that his particular results could have occurred by chance. If this likelihood is very small, the null hypothesis is rejected and the results are said to be significant. If the likelihood that chance was responsible is not very small, the experimenter may be unable to reject the null hypothesis. Under no condition would he accept the null hypothesis.

There are two types of errors which may occur in arriving at conclusions regarding the significance of experimental results. If the independent and dependent variables are really not related but a very rare chance event has made them appear to be, the experimenter may erroneously conclude that a relationship exists. This is referred to as *Type 1 error* and its probability is given by the significance level of the statistical test, *alpha.* If the independent and dependent variables are really related but the experimenter is unable to reject the null hypothesis, this is referred to as a *Type 2 error.*

When variations in the independent variable are associated with small differences in the dependent variable, only an experiment with great *sensitivity* will detect this relationship. Sensitivity increases as the size of the sample increases and as the amount of measurement and experimental error decreases. Since some statistical tests are more capable of detecting differences than are other tests, a property known as the *power* of the test, the sensitivity of an experiment may be increased by using a more powerful statistical test. Another way to increase sensitivity is to attain greater experimental control, by regulating the factors which are known to influence the observed behavior, by employing blocks of experimental subjects who differ less than ordinarily, or by making statistical adjustments of the data.

Beyond increasing sensitivity as much as is economically feasible, the experimenter may attempt to establish the *reliability* of his measurements by clearly identifying and stating the *operational rules* by which they are made. This will increase the *repeatability* of his experiment as well as making it possible to determine whether insignificant findings may have been due to unreliable measurement or weak or nonexistent relationships.

3

The Dependent Variable

In his initial statement of the experimental question, the psychological investigator identifies the general nature of the dependent behavior he plans to observe in his study. As has been suggested in Chapter 2, this statement necessarily aims at a broad, nonspecific level of interrelatability with other relevant research and theory. It is a far cry, however, from verbal expressions such as "strength of conditioning," "degree of anxiety," "amount of retention," etc., to the operationally defined variables that the experimenter may actually end up measuring. The directions which may be taken in generating an operationally defined dependent variable are, literally, almost infinite. Furthermore, much of the philosophy or "style" of his research remains to be characterized by the choices that the experimenter must make in this regard.

This chapter and the next will be concerned primarily with the manner in which dependent variables are selected for use and operationally defined in different experimental situations. The topic is related, on one hand, to the issues considered in Chapter 5, differing largely in the latter's emphasis on the independent variable, and, on the other hand, it is also closely tied to the basic concepts of psychological measurement. Although it may be quite impossible in actual practice to separate these different aspects of the overall topic, it will be useful here to begin with a discussion of the general considerations involved in choosing a particular behavioral measure for study from among the many that are potentially available.

FOCUSING ON OBSERVABLE BEHAVIOR

Experiment on Retention

Suppose that an experimenter wishes to investigate the relationship between the amount of time since the memorization of a poem or a list of nonsense syllables and the degree of retention of the learned material. The actual procedure that will be used to study this relationship empirically depends in substantial part upon the theoretical level at which the experimenter chooses to function. The more theoretical (i.e., explanatory) the experimenter's aim, the more complex his reasoning may have to be in arriving at a choice of observable behavior. But, even the most radical empiricist must resolve many difficult ambiguities of response-definition, whether he gives explicit attention to them or not.

If retention means to the experimenter nothing beyond the probability that the learned response will occur under appropriate stimulational circumstances (or, the proportion of material correctly reproduced), he will be limited to using what is known as the method of *recall.* That is, after the selected amount of time has elapsed since the original learning, the subject will be required to write or otherwise reproduce the original material and some kind of quantitative score will be assigned to his performance. In the case of a list of nonsense syllables, for example, the score might be the number or the proportion of syllables correctly reproduced.

Procedural Decisions

Even at this most simple level, however, significant decisions must be made regarding, for example, how much time to allow for the subject's performance, whether the method of reproduction should be writing or speaking, whether part-scores should be given for partially correct responses, e.g., slightly misspelled nonsense syllables, and the like. The observed difference in performance between subjects who are tested immediately after the original learning and those who are tested many days or weeks later may be influenced by the experimenter's choices in these matters. Subjects who are tested immediately after learning, for example, might be able to reproduce a large proportion of the material in a very short testing period, while those who learned the material a long time prior to testing may need a longer testing period to achieve an equivalent level of performance.

With these considerations in mind, the experimenter might even wish to define response probability, i.e., retention, in terms of time needed, rather than material reproduced, or in terms of some combination of the two, e.g., responses per unit of time. Again, if the subject must reproduce the material orally, he may inadvertently employ the same item more than once. Written reproduction might preclude this type of error, since the subject could compare new responses with those previously written, unless, of course, a decision to prevent this were implemented.

Two subjects whose performances do not differ on a recall task may perform quite differently from one another if a *recognition* measure is used. In a recognition task, the subject must select the originally learned items from a larger list containing nonlearned items as well. The higher the number of correct selections, a score that might require adjustment in relation to incorrect selections, the greater the retention. While the recognition measure differs operationally from the recall measure, it is not a foregone conclusion that one or the other comes closer to the desired meaning of retention. The answer to this question must be made in the context of systematic theory, since it is far from satisfying scientifically to be told that retention means whatever the experimenter wants it to mean.

Actually, the more radically empirical the experimenter (the more disdainful he is of theorizing), the more incumbent upon him it is to demonstrate empirically what the relationship may be between his preferred dependent variable and those used by other experimenters in the same general research area. At an explicitly theoretical level, on the other hand, several different dependent variables may be considered appropriate, with coordinating theoretical concepts employed to account for the observed differences among their empirical relationships with the principal independent variables in the system.

Another potentially useful operationalization of the concept of retention has been employed in many experiments. When particularly long periods of time intervene between the original learning and the retention task, it may be that neither recall nor recognition procedures will be effective without some additional practice. The amount of relearning that is needed to return the subject to his original terminal performance level may also be expected to depend upon and empirically reflect the degree of retention of the original learning. Ebbinghaus' classical experiments on memory, in fact, made considerable use of this method, expressing the subject's retention in terms of savings scores (number of practice trials in original learning-number of practice trials in relearning/number of practice trials in original learning).

The savings score not only defines retention operationally in terms of the amount of relearning that is necessary to restore the original terminal level of performance, it also expresses this measure relative to the subject's original individual rate of learning. It may be noted, again, that more is to be gained from an approach to the study of retention which tolerates (or, even, encourages) the use of all three of the methods mentioned here, as well as cogent others, than from an approach that reduces retention arbitrarily to a single operationally defined measure to the exclusion of the others. This argument seems valid regardless of the degree of experimental control achieved in the study of the arbitrarily chosen dependent measure.

COMPLETELY EMPIRICAL CONCEPTS

Some experimenters would deny, even, that it is fruitful to employ empirically ambiguous concepts of the type exemplified by retention. They would expunge the ambiguity in advance of stating the experimental question, simply by choosing a particular operationally definable dependent variable, such as proportion of correct responses in a stated time period, and indicate that their interest is only in determining the low-level empirical law describing the relationship between time since original learning and this measure.

The criterion on the basis of which the particular dependent variable is selected is said to be nothing more nor less than the degree of control empirically demonstrable between it and various manipulable independent variables. In a science as young and unsystematized as psychology, this approach has the virtue of providing a semblance of orderliness to experimental work. But its essential failure to accept the responsibility of explanatory system development is undeniable. In fact, this approach has sometimes had the unfortunate effect of distorting the basic function of operational definition in psychological science, as some of the critics of operationism have been quick to point out.

The purpose of operational definition is not to make the dependent variable of each and every different psychological experiment a thing apart, having meaning only in the narrowly stated operations by which it is defined in each separate experiment. Its purpose, rather, is to permit the differences among operational meanings to be objectively identified so that coordinating concepts relating them to theory and to one another can be effectively generated and evaluated empirically. When Bridgman (1928) first urged that the meaning of scientific concepts be sought in the concrete operations by which they are measured, permanent fixation of the difference between "length" measured by a straight-edged ruler and "length" measured by the method of triangulation was not necessarily implied; rather, a basis for comprehension of the operational difference in relation to the necessary and desired theoretical articulation between the two concepts was made possible. Only by an operational analysis of the difference between concepts, in relation to an integrative theory, is it possible to understand and even predict the discrepant empirical findings of studies allegedly dealing with the "same" concept.

Retention was chosen for this discussion because its different operational meanings suggest that a broad range of directness–indirectness may be spanned by different studies of the same basic theoretical concept. Consideration of the notion of retention, as it is most frequently employed by the psychologists who have studied it, suggests that it implies some kind of temporal preservation of changes brought about by the original learning. The ultimate operational definition of retention for an experiment of the type mentioned above, of course, cannot be made without some reference to what it is that presumably happens in the original learning. Most psychologists would agree that a neurological process underlies the observable change in behavior potential, although few would be willing to commit themselves to any one neurological possibility. Whatever the underlying neurological change, however, its alteration as a result of whatever happens in time without practice is what results in observable memory loss.

The most direct approach to measuring retention, without commitment to a particular neurological speculation, would appear to be that which is most similar, operationally, to the procedure for producing the behavior change in the original learning situation. In the present example, the retention task should be merely another practice trial, i.e., a recall test. The use of a recall procedure, however, as has been observed above, may sometimes result in retention scores that are too low or too high to permit variations in the underlying process to be reflected, due to so-called "ceiling" or "floor" effects. Thus, some kind of indirect approach may be required. So long as the theoretical concept has implications for observable behavior of the type finally chosen, regardless of the degree of indirectness it exemplifies, an effective empirical test of the experimental hypothesis may be made. At the same time, the tendency toward ingenious indirect dependent measures must be prudently restrained, since careful analysis may often reveal that more direct measures are available.

MOTHER LOVE IN MONKEYS

An especially illuminating illustration of a hierarchy of increasingly indirect dependent measures of the same concept is found in Harlow's (1958) studies of mother love in monkeys. At a general level, the dependent variable of interest in that work was the emotional attachment displayed by a young monkey for its mother, or, as it was most often actually studied, for a surrogate mother made out of wire and sometimes covered by foam rubber and cloth. In the initial effort to define a behavioral implication of this emotional attachment, the relatively direct physical clinging of the infant to the surrogate mother was employed.

With this type of measure, Harlow showed that infant monkeys spend an overwhelming proportion of their time clinging to a cloth surrogate mother, even when a wire surrogate mother is their only source of food. But, the theoretical concept of "love" is no more reflected by this clinging behavior than is the more parsimonious notion that the monkeys preferred the more physically comfortable cloth to the less comfortable wire.

To show that more than mere physical comfort was involved, Harlow devised a situation in which a novel stimulus, which typically evokes a fear reaction in the monkey, could be introduced at a time when the monkey was exploring its environment and both cloth and wire surrogate mothers were physically present. Under these conditions, the monkey almost always reacted to the novel stimulus by quickly retreating from it and then mounting and clinging to the cloth mother. Presumably, Harlow's theoretical concept of love involves the potential employment of the love-object as a safety zone. This result tended to reduce the comparative credibility of the comfort explanation of the previously mentioned preference for clinging to the cloth mother. Yet, it was still possible that the visual transparency of the wire mother rendered it less effective as a hiding place than the cloth mother, since the monkeys did manifest a tendency to hide behind and bury their faces in the cloth mother in the presence of the novel stimulus.

To circumnavigate this explanatory obstacle, a still more indirect measure of the monkeys' love attachment was invented. Under no explicitly engendered motivational deficit, monkeys are known to exhibit a persistent tendency to perform responses which produce visual stimulation, e.g., a monkey will repeatedly operate the lever in a Skinner box to open a window via which a variety of objects may be seen. With an opportunity to view either the cloth or wire mothers in this way, Harlow's monkeys showed a marked preference for viewing the cloth mother, even when the wire mother was their only source of food. This preference for looking at the love-object, manifested in a greater number of barpresses, thus becomes still another observable implication of the concept of love as Harlow conceived it. As is indicated in this example, it may sometimes be necessary to move in the direction of a more indirect operationalization of the basic theoretical concept in order to distinguish it from alternative concepts which imply the same more direct measurement operations.

BACKWARD ACTION

Regardless of the degree of directness–indirectness of the selected dependent variable, certain basic problems of scientific measurement still must be dealt with. Measurement in psychology, as well as in other sciences, may be viewed as being more or less obtrusive upon the fundamental process under investigation (Webb et al., 1966). At the most mechanical level, the tendency for the phenomenon being measured to be influenced by the measurement instrument is known as *backward action.* For example, when the movement of a galvanometer is translated into an ink-mark on a paper recorder, the contact of the pen and ink with the surface of the paper necessarily produces friction. This friction works against the very electrical energy being recorded and tends to reduce or distort its apparent extent. This problem may be dealt with by using low-friction ink, by using electrical pens which "burn" marks on the paper without the necessity of touching it, or even by using relatively frictionless electronic recording devices.

In certain kinds of measurement situations backward action, far from being a problem, may actually be the means by which the measurement is accomplished. For example, if we wish to measure a person's strength by observing how much weight he can lift, the downward action of the weight must be matched by his upward force, exactly to the extent to which this force is available. It is possible to view the effect of the weight he must lift as reducing the person's ability to lift it; the more weight, the less the ability. But, in this case, the backward action involved is the very means by which strength is operationally defined, so there is no problem.

Artificiality of Experimentation

Unfortunately, even the most elegant of frictionless electronic equipment often fails to get at the principal source of backward action in psychological experiments, the artificial confrontation inherent in the experiment itself. An exaggerated illustration of this problem is given by the frequent observation that the interview method for studying purchasing and consumption habits in market research may be more likely to reveal what the respondent views as socially desirable behavior than what he actually does in reality. Asking the average person which paper he reads multiplies the apparent circulation of the *New York Times* by a factor of ten. In these cases, other nonobtrusive measurement devices come to mind readily, although they are typically far more elaborate and expensive to implement.

Occasionally, as has been astutely noted by Underwood (1957), the preliminary features of the measurement procedure may sensitize the subject to be proactively influenced by the independent variable, a special type of backward action whose effect operates "forward" in actual time, rather than retroactively or simultaneously. Underwood was referring to those studies in which a pretest is given to all the subjects, then the different levels of the independent variable are administered, and, finally, the dependent variable is measured in a posttest. In a study, for example, on the effect of viewing a special film dealing with interracial matters upon attitudes toward members of minority races, administration of the pretest before the film may be the enzyme in whose presence the film has its effect, but without which it does not. The problem of sensitization must always be considered when a pretest is employed and appropriate controls introduced to avoid or assess it.

Baseline Measures

A special instance of the sensitization problem may occur in operant conditioning studies in which a standard schedule of reinforcement is employed to establish a baseline "steady state," a more or less unchanging rate of responding. After the steady state is established, a change in experimental conditions is introduced and the effect of this change is evaluated against the background of the ongoing behavior. Since, as Weiner (1964) has shown, the observed alteration in behavior may be as much a function of the particular baseline schedule as the altered conditions, it would be an error to arrive at conclusions regarding the effect of the particular treatment except within the confines of the particular schedule. If the researcher is interested in the possibility that some of the effects of environmental manipulations in experiments of this type may have general significance, it becomes necessary to examine these manipulations under several different baseline conditions. In addition to the possibility that baseline-dependence due to the reinforcement schedule being used may exist, it is also reasonable to expect baseline-dependence due to the actual ongoing response rate and the proprioceptive stimuli this generates.

In a more general way, sensitization may be produced by any of the procedures of an experiment, whether administered prior to the introduction of the independent variable or, even, during the experiment proper. Such conditions as the experimental room, the instructions given to the subject, the experimenter himself, etc., may all contribute to an altered disposition to react to the different treatments in ways uniquely dependent upon all of these procedures. To the extent that the experimenter is on his guard against these sources of contamination, he is frequently able to plan his experiment in such a way that unambiguous interpretation of the results is still possible. Specifically referring to the effects of the dependent variable measurement procedures on the phenomenon under investigation, whether these be mechanical, electrical, physiological, or psychological, the dangers cannot be over-emphasized nor the precautions over-elaborate. It is almost never possible for a new behavioral measure or situation to be employed in an experiment without substantial preliminary study aimed solely at elucidation of the potential avenues of backward action associated with the new measure or situation. For this reason, perhaps more than any other, investigators tend to employ the tried and true behavioral measures of their predecessors.

KINDS OF MEASUREMENT

We should distinguish between the layman's use of the term "measurement" and what we will mean by it in this discussion. Torgerson (1958) has observed that scientific measurement refers more to the conceptual process by which the measurement yardstick is developed for use in a particular experimental setting than it does to the actual use of the yardstick to determine, for example, the length of some object. Our concern is with the manner in which quantitative representation of particular properties of behavior is established.

Even when previously developed measurement procedures are selected for use in testing some new experimental hypotheses, analysis of the logic by which these methods are applied in the new situation is often quite fruitful. This last point becomes particularly cogent when the original establishment of the measurement device involved logic that is demonstrably different from that inherent in the present situation. For example, average magnitude or latency of response, both of which may be useful in studies involving continuously variable responses, may become misleading when the probability of response is what changes in the experiment. A tendency is clearly discernible in the work of many experimenters in psychology to rely upon physical measures, such as time, distance, force, etc., to represent the dependent variables in their experiments. Apparently this is done on the premise that the established mensurational attributes of these physical dimensions will automatically transfer to their situations. The experimenter has the responsibility of showing what psychological or physiological variables are reflected in these physical units and in what way.

Measurement can be classified in two different ways. The first type of classification has to do with the way in which the operations of measurement impart informational properties to the numbers involved; in particular, whether the measurement operations imply *order* only, both *order* and *distance*, or *order, distance,* and *origin* (zero). On the basis of these properties measurement scales are referred to as "ordinal," "interval," or "ratio." Beyond these distinctions, measurement can also be classified on the basis of the inherent logic by which the measurement operations give theoretical meaning to the quantities obtained. Thus, measurement can be either *fundamental,* as when the property lends itself to direct quantitative expression, *derived,* as when the quantities have meaning only in the relationship between two or more other properties fundamentally measured, or entirely *arbitrary,* as when a concept such as "drive" is measured by means of a paper-and-pencil test, or "activation" is measured by pupillary dilation. Our interest in the measurement process in psychological experimentation relates to both of the types of classifications mentioned above, although the more basic scientific issues and problems are found in connection with the latter classificatory scheme (i.e., fundamental, derived, or arbitrary).

Ordinal, Interval, and Ratio Scales

The scalar informational properties of a particular measurement device derive directly from the operational rules under which the actual measurements are made. That is, the common scale of hardness of rocks attained by rubbing each against the others and always scoring the one which scratches the other as "harder" gives operational meaning of a concretely ordinal type to the concept of hardness. What is meant by harder in this instance is completely defined by the concrete operations by which the measurement is made and, by extension, "hardness" is that property of a collection of rocks measured in this way. Similarly, an ordinal scale of "tallness" of men could be established by implementation of the rule that, when two men stand back to back, the one whose head is higher is "taller." Tallness measured in this way means nothing more than the ordinal property defined by the comparative observational procedure.

When an operational rule for determining distance, as well as one for determining order, is added to the measurement situation, the scale obtained will, of course, have equal-interval properties. Again, however, nothing more is meant by the measures than the operational implementation of the rules implies. In the case of tallness, for example, it would be possible to employ some arbitrary unit of distance, such as a small stone or a man's hand, and add up the number of stones or hands needed to go from one end of the measured object to the other. Two marks could be made on a flat surface, corresponding to the two ends of the object in question. Then, the measurement stone or hand could be laid down at one end, such that its end and that of the first mark coincide. A new mark could be made at the other end of the stone and the stone could be lifted and again placed so that one of its ends coincided with the new mark, and so on. The equal-interval properties of this scale of tallness are tied completely and obviously to the procedural rules by which "stones" or "hands" of tallness are added up. In this example, the scale also has an origin, e.g., zero "stones," so that meaningful ratios of tallness can be formed.

However, there is no necessity that the implementation of an additivity rule to obtain an equal-interval measurement scale will automatically also provide information regarding an absolute origin and, accordingly, generate a ratio scale. For example, temperature can be defined by operational rules for the addition of equal proportionate amounts of expansion of the volume of water or some other liquid, as is done in defining the Fahrenheit and Centigrade scales. By virtue of such rules, equal proportionate amounts of expansion of volume are equal temperature intervals and can be marked off accordingly on a thermometer. But, as is true for the two temperature scales in question, the unit's size is arbitrary and an absolute origin has meaning only theoretically when the volume shrinks to the vanishing point. It should be noted that differences between values that have equal-interval scale properties are, themselves, on a ratio scale, a point that will be considered further in the last section of this chapter.

Arbitrary, Derived, and Fundamental Measures

A great deal of measurement in psychological experimentation is of the arbitrary type. This is due partly to the fact that fundamental and derived measurement both depend to a considerable extent upon the existence of systematic theories in relation to which the measurement operations have the status almost of empirical extensions. Although there is nothing inherently wrong with arbitrary measurement, it has a divisive tendency with respect to the goal of systematic theory-construction, since concepts that are measured only by arbitrary fiat have meaning only in relation to their measurement operations.

For example, within the relatively tightly articulated field of research on human classical conditioning, the strength of the tendency to make the conditioned response is measured by eyelid conditioners in terms of the *frequency* or *proportion* of responding over trials, while galvanic skin response conditioners use the *amplitude* or *latency* of the response on each trial for the same purpose. That different empirical laws are obtained in these different research areas is well-known and generally deplored. Yet, little effort has been expended on the theoretical problem of making either or both of these different types of operational measures more than arbitrarily reflective of "response potential." As more experimenters turn toward the problem of developing fundamental measurement, it may be expected that even operationally different measures may begin to have some theoretical meaning in common with one another.

Fundamental measurement requires that quantities can be assigned to represent properties of objects or events according to the laws governing these properties as set forth in theory and without the necessity of measuring any other properties. The operational rules for measuring the length of an object, for example, emanate from physical considerations regarding length, considerations that do not require attention being given to the object's hardness, thickness, or its weight. The operational rules are direct implications of the concept. It is sometimes assumed that only in physical science can measurement be of this fundamental type and produce fully extensive ratio scales. This assumption is based on the more primitive notion that only the empirical addition of quantities (i.e., additivity) can result in a ratio or interval scale of measurement. However, as Torgerson (1958) has shown, fundamental measurement resulting in interval or ratio scales, even in physics, can be based on other fundamental procedures, such as those involved in nonspatial distance relations. Since the latter type of fundamental measurement appears to be quite adaptable to psychological quantification, it will be of interest to review Torgerson's example from physics and its most obvious psychological extension.

A Scale of Temperature

Temperature is ordinarily thought of in physical science as a derived measure, since, as mentioned above, it typically obtains its meaning in relation to other fundamentally measurable properties, such as volume or pressure. To develop a fundamental interval scale measure of temperature, we might first establish an ordinal scale based on the correlation between ordinal variations in temperature and heights of a column of mercury. This could be done simply by heating water in a container and assuming that the higher the column of mercury, the hotter the water. This provides us with a method for identifying and reproducing specimens of water at given unknown temperatures during the procedure of developing our fundamental measure.

It must be clearly understood that we assume no theoretical relation between temperature and height of the mercury column; the latter is to be used solely to identify unknown temperatures, as explained below. We now fill a container to a particular mark with water at temperature *A*, and then pour the water into a larger container. We again fill the small container to the same mark with water at temperature *A*, and heat the water to another temperature, *E*. This is then poured into the same large container. The two quantities of water in the large container are mixed, resulting in a new temperature, *C*. On the basis that we began with two identical quantities of water at temperature *A*, we assume that the obtained temperature, *C*, is halfway between *A* and *E*.

To test this assumption and further develop the scale, we repeat the original procedure using water at temperatures *A* and *C*, to obtain water at temperature *B*, and, again, using water at temperatures *C* and *E*, to obtain water at temperature *D*. Again, it is tentatively assumed that *B* is halfway between *A* and *C* and that *D* is halfway between *C* and *E*. Now, if our reasoning has been sound, and if our theoretical notion that mixture is the appropriate means by which temperature distance may be traversed, we may repeat the procedure, this time using our two new temperatures, *B* and *D*. The resulting temperature should be, of course, *C*, the temperature that resulted from mixing *A* and *E* in the first place.

We may now assign numbers to the five temperatures, such that the lowest number is assigned to *A*, the next to *B*, and finally the highest to *E*, with equal numerical steps between adjacent temperatures. What we will have produced is an equal-interval scale of water temperature based on *division* of *distance* rather than *additivity*. The steps in the scale are equal, since the distance from *A* to *C* equals the distance from *C* to *E*, because *E* bisected the *A–E* interval, and the distance from *A* to *B* equals the distance from *B* to *C* (as well as being equal to the *C–D* and *D–E* intervals), because *B* and *D* bisected the *A–C* and *C–E* distances, respectively. The scale is fundamental in that fundamental measurement of some other attribute was not required for its establishment. Perhaps of even more importance in our discussion, the scale is fundamental in that the basic operational rule for its creation emerges directly from the concept of temperature, i.e., that temperature distance may be traversed by the mixing operation.

Where Did the Scale Originate?

The major difference between the foregoing type of fundamental measurement and the more common arbitrary measurement is that the latter almost always borrows a scaling logic given to it in some other experimental context. Arbitrary measurement may be potentially no more arbitrary than derived or fundamental measurement regarding the use of some objective index, such as blood pressure (measured in millimeters of mercury) to index emotion, in the sense that the theory of emotion may be quite clear in suggesting such an index. The problem, however, is that the theory of emotion says nothing about how many millimeters of mercury are required for equal steps on an emotional scale or even that millimeters of mercury should be the chosen units. When a borrowed index, however, already has its own set of quantitative units, typically of a physical type, the experimenter will seldom undertake the kind of measurement scale development exemplified in the temperature example.

The most serious and sophisticated efforts to construct scales of fundamental measurement in psychology have been undertaken in those research areas in which physical measurement scales are not immediately available for adoption, e.g., in the scaling of psychological attributes of qualitatively different stimuli, in the scaling of preferences and attitudes, etc.[1] In the most behavioristic research areas, however, the tendency has been to use whatever physical units seem most intuitively appropriate (millimeters of eyelid closure, micromhos of conductance-change, latency or duration of response, rate of response, etc.). In many cases such units have been used without any demonstration of the necessary isomorphism between the quantitative units of the physical scale and whatever psychological property is actually under investigation. From the present viewpoint, the casual statement, "The dependent variable is defined operationally as . . . ," would be much less often employed were the potential mensurational inappropriateness of the particular physical scale more carefully weighed.

The psychophysical scaling of loudness, discussed in Chapter 4, will be seen to be very similar to Torgerson's demonstration of distance-scaling of water temperature. In describing the loudness scaling, however, the argument will be made that the theoretical analysis of its principal operational rule appears to have been aimed more at a comprehension of the experience of loudness, per se, rather than the manner in which loudness intervals are divided or multiplied by the subject. Nevertheless, the temperature example may suggest other psychological analogues to imaginative experimenters who are more than cursorily interested in the problem of quantitative expression of the dependent variables they study.

TRANSFORMATIONS OF ARBITRARY MEASURES

The obtained distribution of an arbitrarily defined dependent variable frequently fails to meet the assumptions underlying the most powerful statistical tests that may be made of the hypotheses regarding the effects of the independent variables. This may be due to unavoidable but otherwise trivial limitations of the actual measurement process or it may be due to the actual nonindependence of error and the experimental treatment effects.

Measurement Restrictions

In the first category, for example, the use of proportions as dependent measures may result in binomially distributed measures in which the

[1] The most notable exception to this generalization is found in the psychophysical scaling of physically measurable stimuli, in which physical scales of intensity, frequency, wave length, etc., do exist. Examples from behavioristic studies of learning or motivation, however, are almost nonexistent.

variance in each experimental condition is related to the mean, which is highest when the mean equals 0.5 and increasingly smaller as the mean approaches either 0 or 1.0. In general, this type of nonindependence of the mean and variance may be thought of as resulting from the arbitrary restrictions placed at both ends of the measurement scale, since the proportions cannot be below 0 nor above 1.0. Similarly, a related type of nonindependence of means and variances may result from measurement situations which are limited at only one end of the continuum, by an arbitrary minimum or maximum. In such cases, the mean may actually equal the variance, as in the case of a Poisson distribution.

The use of frequency data sometimes also results in similar limitations on the measurement scale and, in certain situations, the variance of a set of measures may be a function of their number (e.g., ranks). The most common scale changes employed in these cases are the square root and logarithmic transformations. Less frequently used is a reciprocal transformation. When it is desirable that the variance be a function of only n, a special device called the *arcsine* transformation may be used.

Behavioral Effects

In the second category are those situations in which the actual effect of the independent variable may be to increase or decrease the variability of performance in a group of similarly treated subjects whose mean performance is also systematically influenced. In a learning study, for example, the experimental treatment may elevate the performance level of the poorer subjects more than it does that of the better subjects, perhaps because of the initial proximity of the latter subgroup to some kind of physiological or behavioral limit. The consequence will be a reduction in the variability of performance in the experimentally treated group, a reduction which is not observed in a group that received some kind of control treatment. If many of the subjects are at the lower end of the measurement scale at the beginning of a learning experiment, on the other hand, the result of an experimental treatment may be an *increase* of both the mean and variance in the experimental group but of neither in a control group.

Initial Values

Still another source of nonindependence of error and treatment effects is found in the observed positive correlation between initial levels of psychophysiological processes (e.g., heart rate) and the size of observed responses. This problem, referred to as the *law of initial value,* also may result in differential observed treatment effects in subjects whose basal values are higher or lower, particularly, as is often the case, if the experimental treatments tend to influence the basal values as well as the short-term responses.

It is quite common in all of the cases enumerated above for an arbitrary monotonic transformation to be applied to the obtained measures, such that the gross deviations from an optimal distributional form are largely eliminated. The statistical tests are then performed on the transformed values and, in fact, these are often presented in tables and graphs of the experimental results. Sometimes the experimenter may report conventional t and F tests on his data along with another of a variety of so-called nonparametric tests done on the untransformed data. The latter device is a sort of "peace offering" in advance to those of his potential critics who may be wary of transformations or, even, completely intolerant of them.

The basis of this rejection of transformation is quite questionable in the light of the sometimes almost wanton arbitrariness of the original measurement in the first place. There is nothing "holy" about a set of measurements that are not, themselves, fundamentally established. In fact, since many fundamentally measured biological properties (e.g., height, weight) suggest that a normal distribution is the most likely reflection of the variation in natural biological events, the argument may be made that a transformation that achieves normality of distribution moves the measures *closer* to probable biological veridicalness rather than farther away.

Regardless of the persuasiveness of the last point, the argument that the transformed quantities represent some "unknown" measure loses much of its cogency with the recognition that little of a fundamental nature was "known" about the measures before they were transformed. At the very least, it would appear desirable to be able to conduct the most conventional kinds of statistical tests on the data under conditions that approximate their basic assumptions. Recent studies of the effects of violations of the assumptions of the t and F tests testify to the robustness of these tests, suggesting that some of the motivation for transformation of data may have been ill-founded and may be expected to diminish.

Theoretical Factors

Sometimes, transformation of an arbitrary response measure may be suggested by underlying theory. Instructive as a historical case in point is the use of running time as a measure of excitatory potential in simple conditioning studies. Arbitrarily assuming that the time taken by a rat to traverse a straight alley is inversely related to the strength of his tendency to do so (i.e., excitatory potential), the theoretical multiplicative relationship between drive and habit factors in producing excitatory potential militates against the use of an untransformed time measure in studies in which either motivational or associative independent variables are manipulated. If ${}_sE_r = D \times {}_sH_r$, then $\log {}_sE_r = \log D + \log {}_sH_r$. Using a logarithmic transformation of the time measure, thus, permits an *additive* combination of the motivational and associative factors to operate. If one of these is held constant in the experiment, logarithmic (or geometrical) variation in the other should be linearly related to the logarithm of the dependent time measures.

The transformation is both appropriate and useful in testing these implications of the theory. It also has the virtue of reducing positive skewness of the obtained distributions. The logical appeal of this argument notwithstanding, many investigators in this research area persist in using a *reciprocal* transformation of time measures, frequently using the distance of the runway in the numerator. The resulting unit is, of course, a *speed* measure, typically expressed in feet per second. Since the reciprocal transformation also tends to reduce positive skewness (as well as benefiting from the "face validity" of the positive speed-excitation relationship), there is little reason to expect either of these methods to be totally supplanted by the other. An apparently untried compromise between the two methods, the logarithm of speed, would also appear to have promise in this field.

A somewhat different instance of transformation of a more or less arbitrary response measure, suggested by psychological theory, is found in two recent studies (E. Kimmel, 1967; Pendergrass and H. D. Kimmel, 1968). In these studies, subjects judged the intensity of either an electric shock or a pure tone by rotating a plastic lever through an angular distance that was proportional to their subjective impressions of the intensity. Since Stevens (1957) has found that magnitude estimations of this type are power functions of physical stimulus intensities, a logarithmic transformation of the actual response magnitudes (originally measured as degrees of rotation of the lever) would be expected to produce a variable which is a simple linear function of the logarithm of physical intensity. Even though stimulus intensity was not itself varied in these studies, the logarithmic transformation was deemed to be appropriate. As is noted in Chapter 5, geometric or logarithmic manipulation of even *independent* variables may be equally implied by theory and used to advantage in certain research areas.

Difference Scores

It was mentioned earlier in the chapter that *differences* between interval scale values have ratio scale properties. Since psychological investigators frequently are in a position to transform their dependent measures to difference-scores, using some preliminary measure of the dependent behavior as either the minuend or subtrahend, depending upon which measure is larger (so that positive remainders will be the rule), the fact that the resulting quantities may have ratio scale properties is an important added benefit. Not only does the difference-score tend to reduce error from individual differences among the subjects (which is its main purpose, as discussed more fully in Chapter 7, p. 144), in addition, it may enable the experimenter to avail himself of the extensive measurement properties of ratio scale measures. This would include both the opportunity to use statistical methods that necessitate ratio scale numbers (Senders, 1958) as well as the possibility of arriving at empirical laws with important quantitative theoretical properties. Since this implication of the difference-score transformation has not previously been emphasized it may be that researchers will be more inclined than they have been in the past to use it when feasible.

Quickening

Transformation of the dependent variable sometimes is done automatically during actual data-collection (i.e., in "real" time) for the purpose of providing the experimenter with an immediate rendition of some particular version of the data in which he is most interested. In the field of engineering psychology this is known as *quickening*, a term which refers to the direct presentation of a mathematical function of quantitative information rather than the raw (unquickened) information itself. For example, an electronic display might present real-time fluctuations in the distance between an attacking ship and the target, rather than separate presentations of the actual locations of the two. Of course, the unquickened presentation of the locations of the two objects would permit later determination of the distance between them, but, if the distance between them was of critical importance in guiding the attack, it would be important to get this information immediately rather than having to await computations after the fact.

In basic psychological research, a good example of direct transformation of the dependent variable in real time occurs in the use of the cumulative recorder in operant conditioning research. Because this type of recorder has a constant paper speed, and because the recorder-pen moves a standard distance upward for each response and remains there until the next response moves it upward again, the slope of the recorder line provides a direct presentation of the *rate* of responding. Since many experimenters in this area of research have stressed rate of responding as a dependent variable (rather than, say, number of responses, or strength of response), the real-time transformation is useful to them.

SUMMARY

Problem analysis refers to the manner in which the experimenter proceeds from the general statement of the experimental question to the operationally defined variables actually employed. Of primary importance in problem analysis is the matter of selecting the dependent variable. It is here that the experimenter finally focuses on observable behavior. His actual choice among possible contenders will naturally depend upon the theoretical level at which he functions. Sometimes it is possible to broaden the implications of the study by using two or more different operational specifications of dependent behavior. For example, *retention* could be defined in terms of *recall, recognition,* or *savings* scores. Obviously, depending upon one's views regarding what it is that presumably is learned, different methods for measuring its retention would be appropriate. In certain situations it may be necessary to use rather *indirect* measures of the dependent variable, when direct measures are unavailable.

In almost any measurement situation the phenomenon being measured may be influenced by the measurement operations. This is known as *backward action* and may often be a problem. In psychological experimentation, the most ubiquitous source of backward action may be the artificial confrontation of the experimental situation. The obtrusiveness of the experimental situation should of course be kept to a minimum.

Scientific measurement may be classified in two different ways, one dealing with the informational properties acquired by the measures as a result of the operations by means of which they are obtained (e.g., *ordinal, interval,* and *ratio* scales) and the other depending upon the inherent logic by which the measures gain theoretical meaning (e.g., *fundamental, derived,* or *arbitrary*). The first classification depends entirely upon the operational rules by which the numbers are assigned. The *interval scale* of measurement is seldom surpassed in psychological research and probably not often enough achieved. Likewise, most psychological measurement is of the *arbitrary* type, although the possibility of fundamental measurement in psychology is not precluded. Because most psychological dependent variables involve arbitrary measurement, *monotonic transformations* are often used to enable preferred statistical procedures to be used or to provide *quickened* quantitative information to the experimenter in real-time. A common example of this in psychology is the use of a cumulative recorder to display *response rate* directly.

4

Psychophysics and Scaling

PSYCHOPHYSICAL DEPENDENT VARIABLES

In the early days of scientific psychology, prior to the behavioristic revolution of the first two decades of this century, psychology was identified as the science of the mind, or of mental activity. The first major scientific assaults against the barriers of psychological ignorance were aimed especially at the ways in which sensory impressions depend upon variations in external physical events. The human subjects in these studies were viewed as instruments of detection of stimulus changes. In this context, it was desirable to reduce the vast potential for response variation to its most reliable minimum, ideally to a dichotomy, and to engage in systematic variation of the stimulus material while the response went on and off. Thus, while the main body of psychophysical research consisted of stimulus manipulations and response observation, the dependent variables that emerged from this work *were expressed as stimulus quantities,* with only rare exception. Given that the subject was being used as a stimulus detection instrument, it is not surprising that stimulus units were employed to describe the experimental results.

Method of Limits

For example, in using Fechner's classical *Method of Limits* to determine either the absolute or difference thresholds, the subject's response options are limited to either a dichotomous "yes" or "no," in the first case, or to the equally dichotomous "greater" or "lesser," in the second. The *Method of Limits* involves a step-by-step process of stimulus change, either ascending from below the threshold to above it, or descending from above the threshold to below it. Each time the "stimulus" is presented, the subject must indicate whether or not he detects its presence (in measuring the absolute threshold), or he must indicate whether the comparison or standard stimulus is greater (in measuring the difference threshold). The subject is not permitted to report that the stimulus is "very much" present, or "very little" present, or that it is "barely" absent. He is reduced literally to a "go-no-go" indicator, a flip-flop.

Quantification of the dependent variable is achieved in psychophysics by inference from *how often* a stimulus of a given value is detected, or *how often* a given difference is detected. The manner in which this inference is made is usually quite complex and involves a variety of basic assumptions regarding the stimulus–response process. In this manner it illustrates the way in which the measurement operations are empirical extensions of basic theory. It also exemplifies a number of quantitative procedures that occur repeatedly in both psychophysical and other types of experimental research. It is convenient to describe the main features of this complex procedure in the context of another of Fechner's classical methods, the *Method of Constant Stimuli,* since from a methodological point of view, it was a distinct improvement over the older *Method of Limits.*

Constant Stimuli

The Method of Constant Stimuli was devised to eliminate the main methodological problems inherent in the *Method of Limits,* problems that result from the systematic fashion in which the stimulus variation is accomplished. As a result of repeated experiences with either ascending or descending trials, in the *Method of Limits,* the subject frequently may develop a tendency either to become so accustomed to the particular response he has been making (e.g., "I don't see it"), that he continues to make the same response even after the threshold is actually crossed and the response should be changed, or he may attempt to anticipate the stimulus changes in advance of their occurrence. The first of these tendencies is referred to as the *habituation* error and the second is known as the *expectation* error. In spite of the fact that they tend to work in opposite directions, it is usually impossible to determine in a particular experimental situation which, if either, may predominate.

Determining the Absolute Threshold

The *Method of Constant Stimuli* employs a fixed set of stimuli, chosen [1] to range in equal physical steps on either side of the threshold, so that

[1] The actual stimuli are chosen on the basis of prior information, frequently obtained using the *Method of Limits.*

the smallest would *almost* (but not quite) *never* be detected while the largest would *almost* (but not quite) *always* be detected. The schedule of delivery of these stimuli is completely unsystematic, so that the subject can never either get into the habit of making a particular response nor anticipate a change of a predictable type. Each stimulus value is presented many times. Thus, the proportion of times it is detected can be determined relatively reliably. These proportions represent the starting point of the complex process of threshold determination. For illustrative purposes, a hypothetical set of proportions based on the behavior of a single subject whose absolute intensity threshold for acoustic energy (using a 1,000 Hz tone) was measured by the *Method of Constant Stimuli* is presented in Figure 4.1. The curve connecting the data-points is known as a *psychophysical function.*

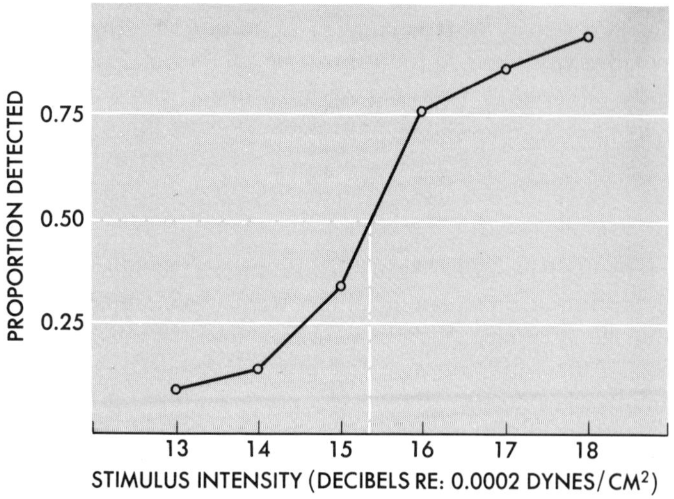

Figure 4.1 Psychophysical function obtained in hypothetical study of the absolute intensity threshold for acoustic energy, using the *Method of Constant Stimuli.*

The absolute threshold is defined as that stimulus intensity which the subject notices half of the time. This is an arbitrary, statistical definition which illustrates the fact that considerable fluctuation from occasion to occasion actually occurs in response to stimuli in the vicinity of the threshold. A simple empirical demonstration of this fluctuation can be achieved by slowly moving a ticking watch away from the ear until its sound appears just to vanish. Holding the watch fixed at that point, it will be noticed that the sound seems to wax into and wane out of psychological existence. Defining the threshold in statistical terms takes these observable fluctuations into realistic account.

The psychophysical function is frequently bisected symmetrically at the 0.5 point, in the sense that the arithmetic mean and the median are usually very close to each other. Thus, the statistically defined threshold is also the weighted average value of the stimuli concerning whose presence the subject is in doubt.

Linear Interpolation of the Threshold

The easiest and most obvious way in which to estimate the value of the threshold from the data of Figure 4.1 would be simply to draw a horizontal line across the graph, parallel to the abscissa, from the 0.5 point on the ordinate to the place where it intersects the psychophysical function. Then, a perpendicular may be dropped to the abscissa to determine the stimulus value associated with 0.5 detection. These lines are also shown in Figure 4.1, indicating that the absolute threshold in this example is slightly below 15.5 decibels.

There are two basic deficiencies inherent in this method of linear interpolation. First, and more important, data from only two of the stimuli are actually used to determine the threshold. These are the stimuli which were detected just below and just above 0.5 of the time. In fact, if one knew with reasonable certainty that the data would turn out as they did in this hypothetical example, it would have been much easier simply to use the 15 and 16 decibel stimuli only. Consideration of this point also should reveal the fact that the actual selected values of these critical straddling stimuli may have a disproportionately great influence on what the obtained value of the threshold turns out to be, since the manner in which they straddle the 0.5 point will seldom be exactly symmetrical. The second deficiency of the method of simple linear interpolation is that it assumes that the function *is linear* between the two middle values. Although this assumption is not usually very far from being accurate, it fails to allow for whatever curvilinearity may occasionally be present.

Phi–gamma Assumption

The method developed by the classical psychophysicist to deal with these problems begins with an extremely significant theoretical assumption, namely that the shape of the distribution of judgmental responses around the threshold value is *normal*. This assumption, known traditionally as the *phi–gamma* assumption, is based on the observed fact that a chance-like process seems to be operating in the zone of transition from subthreshold to suprathreshold stimulus values. The reason that the psychophysical function does not *appear* to have the well-known shape of the normal distribution is quite simple; it is, as it should be, a normal *ogive*, a cumulative normal distribution. That is, as the intensity of the stimulus is increased, an increasing accumulation of response tendency builds up. Inspection of the psychophysical function in Figure 4.1 demonstrates its close resemblance to the cumulative normal curve, a point which is the basis for empirical testing of the phi–gamma assumption in particular experimental settings.

Once the phi–gamma assumption has been made, it is then quite easy to transform the obtained proportions to values of normal–deviates (z values, obtained from tables of the normal distribution), above and to the left of which are cut off areas under the curve equal to these cumulative proportions. That is, a proportion of 0.5 would be transformed to a z of 0, since 0 is the midpoint of the normal distribution and half of its area lies above and to the left of it. Similarly, a proportion of 0.84 would be transformed to a z of 1.0, since a perpendicular erected at $z = 1.0$ cuts off 0.84 of the area under the curve (it will be recalled that approximately 0.68 of the area under the normal curve lies between z values of ± 1.0, 0.34 on each side of $z = 0$). A proportion of 0.16 would be transformed to a z of -1.0, by the same line of reasoning. All proportions below 0.5, of course, are transformed to negative z's and all proportions above 0.5 are transformed to positive z's. Figure 4.2 shows the transformed psychophysical function derived on the basis of the normal–

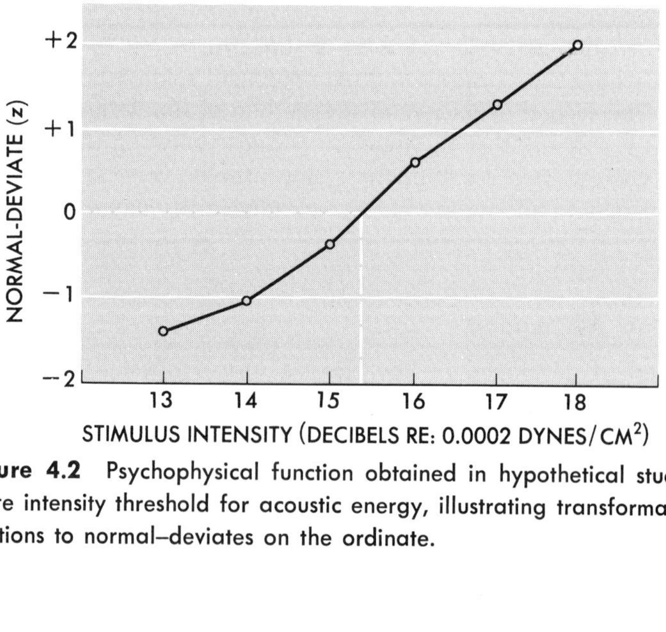

Figure 4.2 Psychophysical function obtained in hypothetical study of the absolute intensity threshold for acoustic energy, illustrating transformation from proportions to normal–deviates on the ordinate.

deviate equivalents of the proportions of Figure 4.1. Again, straight lines are used to connect the points in the figure.

It is clear in Figure 4.2 that the phi–gamma assumption in the present situation receives substantial support (in our hypothetical data), since the points based on the normal distribution transformation fall extremely close to a straight line. The fact that they do not *exactly* describe a straight line is taken merely to mean that some unreliability remains in the situation, which, if completely eliminated, would result in a perfectly linear transformed psychophysical function. Again, linear interpolation could be used to estimate the threshold, as has been done in the figure. But, as was true previously, this method fails to use all of the data available in the situation (even though, in this case, all of the data *are* used to support the phi–gamma assumption).

Method of Least Squares

On the assumption that a straight line would best describe these data, aside from the slight unreliability shown in some of the points, a method is available for fitting a straight line to the actual data-points which uses all of them. This method, known as the *method of least squares,* determines the equation of the straight line above and below which the sum of the squared deviations of the actual data-points is at a minimum.[2] This line is a sort of "running average" of all of the data-points in the experiment, in the same way that the arithmetic mean of a set of scores is the value around which the sum of the squared deviations is at a minimum. The slope of the least squares line, b, is determined from the stimulus values (S) and the normal–deviates (z) in standard fashion, using the following equation:

$$b = \frac{n\Sigma Sz - (\Sigma S)(\Sigma z)}{n\Sigma S^2 - (\Sigma S)^2}$$

The value of the absolute threshold is also determined by formula, simply by first determining the y intercept, a, by substituting into the formula $a = \overline{X}_z - b\overline{X}_S$, and then solving the equation of the line for the value of S when $z = 0$ (i.e., the threshold equals $-a/b$). It is not necessary actually to draw the least squares line, although for illustrative purposes this has been done in Figure 4.3 on the basis of the computations shown in Table 4.1. As is indicated in Figure 4.3, the least squares line does not touch all of the obtained data-points, but it comes extremely close to all of

[2] The general equation of a straight line, using z in place of y and S in place of x, is $z = a + bS$. The slope of the line, b, gives the rate at which z (the subject's reaction) changes relative to changes in S (the stimuli), while its y intercept (the value of y when $x = 0$) is given by a.

them. Again, the determination of the threshold by graphic interpolation is shown in Figure 4.3; however, in this case all of the points *were* used in determining the location of the line (so that the graphically determined threshold agrees quite closely with the value of 15.18 decibels computed by formula in Table 4.1).

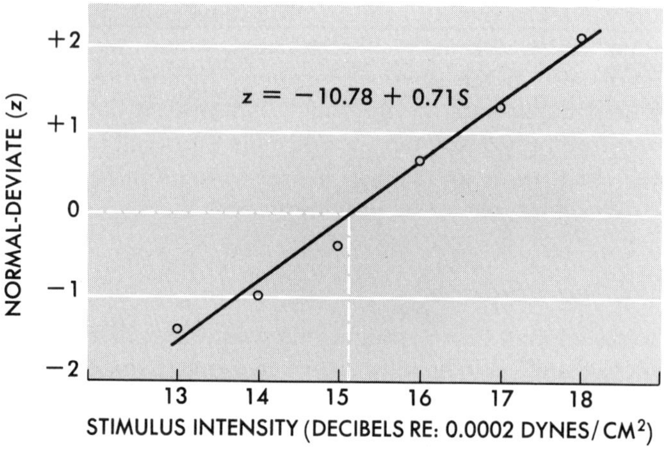

Figure 4.3 Psychophysical function fitted by the method of least squares. The equation written on the graph was determined by formula as shown in Table 4.1.

Determining the Difference Threshold

Figure 4.4 shows another set of hypothetical data, this time illustrating the way in which the *Method of Constant Stimuli* may be used to determine a difference threshold (DL), again for intensity of acoustic energy. In this case a 50 decibel, 1,000 Hz tone was the standard stimulus and eight comparison stimuli were used, ranging in 1 decibel steps from 46.5 through 53.5 decibels. Each comparison stimulus was presented either just before or just after the standard stimulus and the subject indicated which was louder (he was not permitted to say "equal"). The subject was uninformed regarding which stimulus was the standard and which the comparison. Each of the eight comparison stimuli preceded the standard half of the times it was presented and each followed it half of the times it was presented. Figure 4.4 shows the proportion of times that each comparison stimulus was reported to be *louder* than the standard. As is indicated in the figure, a normal ogive again seems to be described by the obtained proportions.

Applying the same logic employed in determining the absolute threshold, the proportions are transformed to normal–deviates, as is shown in

TABLE 4.1

Computation of the absolute threshold (AL) and equation of the least squares line for the data shown in Figures 4.1, 4.2, and 4.3.

S	P	z	Sz
18	0.98	2.06	37.08
17	0.88	1.18	20.06
16	0.79	0.81	12.96
15	0.35	-0.39	-5.85
14	0.16	-1.00	-14.00
13	0.09	-1.35	-17.55
Σ = 93	–	1.31	32.70

$$n = 6$$

$$\Sigma S^2 = 1459$$

$$b = \frac{(6)(32.70) - (93)(1.31)}{(6)(1459) - (93)^2}$$

$$b = 0.71$$

$$a = 0.22 - (0.71)(15.5)$$

$$a = -10.78$$

$$\text{AL} = -\left[\frac{(-10.78)}{0.71}\right]$$

$$\text{AL} = 15.18$$

Least squares equation: $z = -10.78 + 0.71S$

Figure 4.5. Also shown in Figure 4.5 is the least squares line fitted to these data according to the calculations shown in Table 4.2. Again, the phi–gamma assumption appears to have been justified, since the least squares line comes very close to all of the hypothetical obtained data-points.

Point of Subjective Equality

It will be noted that *two* horizontal lines have been extended in Figure 4.5 from the z ordinate to the points at which they intersect the psycho-physical function. One of these lines originates at $z = 0$, while the other originates at $z = 1.0$. A moment's thought will make it clear that the difference threshold cannot be obtained in the same fashion as was used

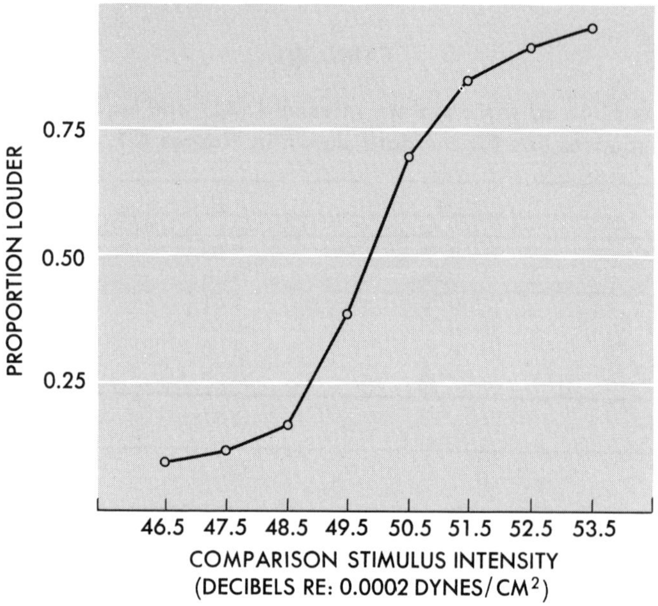

Figure 4.4 Psychophysical function obtained in hypothetical study of the intensity difference threshold for acoustic energy, using the *Method of Constant Stimuli.*

to determine the absolute threshold in our previous example. Dropping a perpendicular from the point on the psychophysical function corresponding to a z of 0 gives us the value of the comparison stimulus that was reported (or, would have been reported, had it actually been used in the study) to be louder than the standard *half* of the time (i.e., $p = 0.5$). This value is known as the *point of subjective equality* (PSE), on the assumption that a stimulus which is reported as louder than the standard half of the time (and, therefore, as softer than the standard half of the time) may properly be referred to as subjectively *equal* in loudness to the standard. But, if the 0.5 response point gives us the PSE, what is the *threshold* in this situation?

Just as it was necessary in the case of the absolute threshold to adopt a statistical definition, a somewhat arbitrary statistical criterion must also be used in defining the difference threshold. The amount of difference between the standard and comparison stimuli that was reported as "louder" 50% of the time will obviously *not* do, since the subject always identifies one of the stimuli as "louder" by instruction and would achieve the 50% level simply by guessing. What is needed is a criterion based upon some proportion greater than 0.5. But, how much more often than half of the time must the subject report that a stimulus is louder than

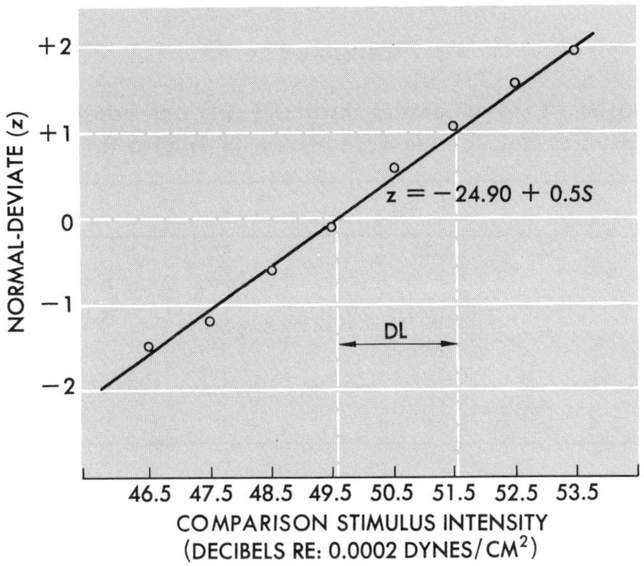

Figure 4.5 Psychophysical function fitted by the method of least squares, according to the calculations given in Table 4.2. The PSE is determined graphically in the figure by extending a line from $z = 0$ to the function and dropping a perpendicular. The DL is determined graphically by extending a second line from $z = 1.0$ and dropping another perpendicular. The DL is the distance on the abscissa between these two perpendiculars.

the standard for us to consider the difference to be of threshold magnitude?

The traditional solution to this problem was to demand "half-again more than half," or, in other words, to require that a difference be judged as louder 0.75 of the time. Thus, a threshold difference is traditionally defined as one which results in the stimuli being reported as different (or one of them as louder) 0.75 of the time. The 0.75 criterion was also thought to be convenient because the middle 50% of the normal distribution (from the 0.25 point to the 0.75 point) was commonly used in statistical inference as the *probable error.*

The Standard Deviation as the Threshold

However, with the increased sophistication of more recent mathematical statistics, the probable error has gone out of fashion and the more rational standard deviation (and standard error) has come into universal use. Since the ordinate of our psychophysical function is already in standard deviation units (z is the standard deviation of the normal dis-

TABLE 4.2

Computation of the difference threshold (DL) and equation of the least squares line for the data shown in Figures 4.4 and 4.5.

S	P	z	Sz
53.5	0.96	1.75	93.62
52.5	0.90	1.28	67.20
51.5	0.84	0.99	50.98
50.5	0.72	0.58	29.29
49.5	0.40	–0.25	–12.38
48.5	0.22	–0.77	–37.34
47.5	0.11	–1.23	–58.42
46.5	0.06	–1.56	–72.54
Σ = 400	–	0.79	60.41

$$n = 8$$

$$\Sigma S^2 = 20{,}042$$

$$b = \frac{(8)(60.41) - (400)(0.79)}{(8)(20042) - (400)^2}$$

$$b = 0.50$$

$$a = 0.10 - (0.50)(50.0)$$

$$a = -24.90$$

Least squares equation: $z = -24.90 + 0.50S$

PSE: $(0 = -24.90 + 0.50S) = 49.80$

DL $= 1/b = 2.00$

tribution), an even more convenient (and less arbitrary) measure of the difference threshold becomes available, namely, the standard deviation of the psychophysical function. Thus, in the present example, the difference threshold is that amount of difference in physical stimulation that is necessary to shift the subject's response from $z = 0$ to $z = 1.0$, i.e., one standard deviation on the transformed response scale. In Figure 4.5 this is the distance on the abscissa between the two perpendiculars dropped from the psychophysical function, as is indicated.

Having defined the difference threshold as that amount of change in stimulation that is necessary to produce one standard unit of change in response (i.e., from $z = 0$ to $z = 1.0$), another convenient feature of this method of defining the threshold is revealed. It will be recalled that the slope of the psychophysical function, b, describes the rate of change in z relative to change in S. Since the threshold is defined as that amount of change in S associated with a unitary change in z, it turns out that the threshold is the reciprocal of b, so that we may simply invert our previously given formula to compute it:

$$\mathrm{DL} = \frac{n\Sigma S^2 - (\Sigma S)^2}{n\Sigma Sz - (\Sigma S)(\Sigma z)}$$

This relationship is more than computationally advantageous. Given that the slope reflects the rate of change in response as the intensity of the comparison stimulus increases, it should be apparent that an individual with fine discriminative ability (one with a *small* threshold) *should* have a psychophysical function with a steep slope, since his responses should change very rapidly as the stimulus is changed. A poor discriminator (one with a large DL) *should* have a psychophysical function with a very flat slope, since there would be little difference in how often he judged the comparison stimulus as louder over the range of intensities used. Since the DL is taken to be the reciprocal of the slope, the fine discriminator (with a steep slope) would have a small threshold, while the poor discriminator (with a flat slope) would have a large threshold. Viewed in relation to the theoretical concept of discrimination, the slope of the psychophysical function is a fundamental measure of its empirical implications (i.e., the rate of change in response relative to stimulus change). The reciprocal is equally fundamental, albeit styled to produce *stimulus* rather than *response* quantities.

A final note is in order regarding the difference threshold defined as the standard deviation of the psychophysical function. As has been noted above, it was conventional in the past to describe the threshold in terms of a response criterion expressed in terms of the proportion of the time that the subject reports that the comparison stimulus is louder (or otherwise different) from the standard stimulus.

Can we express the response criterion associated with the standard deviation in analogous terminology? Put another way, the question is, "What proportion of responses is involved when the standard deviation of the psychophysical function is the criterion of definition of the threshold?"

The answer to this question, while perhaps not immediately obvious, is actually quite simple. Given that the *phi–gamma* assumption is valid and, thus, that the subject's discriminative responses *are* normally distributed as a function of stimulus values in the range of uncertainty, it may be said that the difference threshold is that amount of difference between the standard and comparison stimuli that is detected 0.84 of the time (compared with the classical 0.75). We arrive at this value by noting that the area under the normal curve above and to the left of a z of 1.0 equals 0.84 of the total area under the curve (i.e., 0.5 up to a $z = 0$ plus 0.34 from $z = 0$ to $z = 1.0$). Thus, to the empirically assessable extent to which the least squares line is accurately descriptive of the plot of z on S, 0.84 is the response criterion achieved by the use of the standard deviation of the psychophysical function as the difference threshold. Since serious deviations from goodness of fit of the least squares line would tend to vitiate the entire logic of the method anyway, it is reasonable to describe the 0.84 value as applicable in general to those cases in which the method is justifiably used. It may be expected therefore that the standard deviation method will result in values of the difference threshold that are generally slightly larger than those obtained by the older, 0.75 criterion.

It should be noted that an experiment on the effect of some independent variable on the absolute or difference thresholds would involve as many repetitions of one of these complex procedures just described as there are subjects in the study. The entire procedure of presenting all of the stimuli repeatedly, transforming the proportions of responses to normal–deviates, fitting a linear equation by the method of least squares, and, finally, determining the value of the threshold must be done separately for each subject. The result of this very tedious and complex process is a single value of the dependent variable for each subject. It would be less than accurate to suggest that psychophysical dependent variables are typical, in this respect, of the majority of dependent variables in psychological experimentation; yet, it *is* true that they are representative of the *kinds* of complexities that often may be involved.

Response-Variation in Psychophysics

In addition to the methods previously mentioned, Fechner was responsible for a third classical psychophysical method, the *Method of Average Error*. This method holds an intermediate position between the stimulus-oriented devices for estimating thresholds (i.e., *Limits* and *Constant Stimuli*) and the more modern methods of behavioristic psychology which typically entail unlimited response-variation.

The *Method of Average Error* involves the presentation of a standard stimulus and a variable stimulus, but provides for variation of the latter *to be accomplished by the subject*. For example, the subject is given the task of manipulating the value of the variable stimulus (usually manually by means of some kind of mechanical or electrical control) until he is satisfied that it is subjectively equal to the standard. This method has been very useful in the quantitative specification of perceptual illusions, such as the vertical–horizontal illusion and Müller-Lyer illusion, as well as being used to study the influence of various manipulable associative and motivational factors on quantitatively specifiable perceptual responses (e.g., the effect of value on the perceived size of objects, discussed in Chapter 5).

Constant and Variable Errors

The reason for describing the *Method of Average Error* as a bridge between the stimulus-oriented psychophysical methods and the response-oriented behavioristic methods is that it shares some of the characteristics of each. On one hand, the subject is given complete "freedom" to adjust the variable stimulus to any value in its manipulable range (i.e., the responses are *not* limited to a small number of categories). But, on the other hand, he *is* limited to the task of matching the standard stimulus; and, in addition, the dependent variable again is expressed in stimulus units. The reason for the name, *Average Error,* is simply that Fechner assumed that a large number of trials would permit unsystematic *variable* errors to cancel out, since these errors are generally thought to result from random factors. After the *variable* errors are, thus, eliminated from consideration (except, occasionally with respect to their extent, in the form of a standard deviation), any remaining difference between the value of the standard stimulus and the average setting of the variable stimulus is a measure of the *constant* effect being investigated (e.g., the illusion). As has been noted above, this result is expressed typically in stimulus quantities.

SCALING OF STIMULI

Two somewhat different goals are identifiable in the various efforts to develop psychological measurement scales. The major emphasis has been on the establishment of interval or ratio scales of psychological stimuli (retaining the traditional aim of psychophysics), based on subjects' perceptual reactions to the stimuli. Frequently, the scale values may then be used to define the *independent* variables of stimulus–response research (e.g., Hovland's work, 1937, on stimulus generalization, using JND scale of stimuli). Another goal of scaling, which admittedly is barely developed in actual practice, is that of specifying psychological dependent variables on the basis of arbitrarily measured physical response indices (e.g., the work of Hull, Felsinger, Gladstone, and Yamaguchi, 1947, which attempted to develop an equal interval scale of habit strength). It would appear that a great deal is to be gained from more frequent consideration of the psychological scalar properties of subjects' physically measured responses.

Within the general area of scaling of stimuli on the basis of subjects' perceptual reactions to them, two rather different approaches have been taken. The first, exemplified by Thurstone's work (1927), is a logical outgrowth of the thinking developed in classical psychophysics. It is based on the now familiar assumption that subjects' discriminative reactions are normally distributed in relation to eliciting stimulus events. The second approach, characterized best by Stevens' (1961) ratio scaling methods reflects a radical departure from the traditional methods. In the terminology of the first section of this chapter, Stevens' methods are based on the assumption that the dimensions of psychological reactions to stimuli (such as the perceived loudness of sounds) can be measured fundamentally in a direct way, simply by having the subject add his experiences together or divide or multiply them. Both Thurstone's and Stevens' approaches provide empirical means for evaluating the internal consistency of the scale values obtained, similar in principle to that which was used to check on the temperature scaling method described in Chapter 3.

Scaling from Paired Comparisons

Suppose that an experimenter wishes to study the hypothesis that the esthetic response to a painting or a sketch (i.e., how beautiful it is reported to be) depends upon the extent to which it is an objectively accurate rendition of its subject matter (confining himself, of course, to pictures that actually *have* explicitly identifiable subjects, i.e., *representational art*). Even if it were possible for the same artist to make drawings of the same objective subject matter, intending to exhibit varying degrees of representational accuracy, it would be no mean task to develop a quantitative scale of this property of the sketches, a scale on which they could be located in relation to one another. Of course, for the purpose of testing the hypothesis at its simplest level, it would only be necessary to identify *two* examples which differed quite obviously in degree of representativeness. Then, some type of esthetic response to the sketches could be obtained from subjects and it would be possible to determine whether its average value differed as a function of representativeness.[3]

However, if the proposed research aims at a quantitatively more sophisticated target (e.g., as would be the case if the experimenter wishes

[3] The present section deals with the problem of scaling the *stimuli* with respect to their degree of representativeness. In a later section in this chapter, the problem of scaling subjects' reactions will be considered. Since, as is noted above, the stimulus dimension of representativeness is scaled on the basis of subjects' responses, it will come as no surprise that the same operations may be involved in both cases.

to obtain an estimate of the *quantitative relationship between degree of representativeness and strength of esthetic response*), more careful consideration would have to be given to the method by which representativeness was to be measured in the first place. It is clear that the definition of the concept involves, not some physically specifiable dimension, but one that must be inferred from the reactions of human subjects. Given two sketches of the same subject matter, which one *seems* (which *looks*, etc.) to be more representative of the subject matter? If the typical individual can answer a question of this type, expressing a simple discriminative reaction, it will be possible to develop an equal-interval scale of the dimension in question.

The *Method of Paired Comparisons* typically involves a set of stimuli, ranging in number from as few as four or five up to as many as eleven or twelve (more than about twelve stimuli would generate an experimental task that might be too boring or fatiguing to the subject, since $n(n-1)/2$ comparisons must be made when there are n stimuli). The stimuli are presented in pairs (e.g., in the case of the artistic sketches, the pairs would be presented *simultaneously*, but, for certain types of stimuli, such as auditory stimuli, *successive* presentation is necessary), and the subject is required to make some kind of discriminative response (e.g., better, taller, more attractive, or in the present case, *more representative*) to one of the members of each pair. Each stimulus is paired several times with each of the others, so that it is possible to obtain a proportion of preferences for each *vis a vis* all the others.

The data thus collected are assembled in a proportion matrix, as is shown for a set of ten artistic sketches in Table 4.3. In collecting the data shown in Table 4.3, a photograph of the actual subject matter used by the artist was available to the subject as he examined each pair of sketches.[4] The proportions shown in the cells of the table are based upon the number of times each sketch listed at the left was chosen as being more representative than the sketch listed at the top. Thus, for example, sketch A was judged to be more representative than sketch B, 0.67 of the times they were compared; sketch C was judged to be more representative than sketch F, 0.83 of the times they were compared, etc. A proportion of 0.50 is entered in those cells in the table where a sketch is compared to itself. The extreme right-hand column of the table presents the sums of the proportions of choice for the sketches listed at the left. As can be seen from inspection of these values, the sketches have been arranged so that the sketch with the highest sum is listed at the top, etc.

[4] The subject matter actually used in this study consisted of three vases, assembled on a table. The artists were instructed to use these objects in a drawing, but to arrange them in any way they wished.

TABLE 4.3

Proportion matrix for paired comparison scaling of *representativeness* of ten artistic sketches. The sketches are listed in order of the summed proportions of preference, given in the extreme right-hand column.

Sketches		E	G	I	D	H	F	A	C	J	B	Σ
	E	0.50	0.67	0.72	0.78	0.83	0.83	0.78	0.83	0.89	0.83	7.66
	G	0.33	0.50	0.50	0.61	0.72	0.78	0.78	0.72	0.94	0.94	6.82
	I	0.28	0.50	0.50	0.61	0.72	0.72	0.67	0.61	0.89	0.89	6.39
	D	0.22	0.39	0.39	0.50	0.67	0.67	0.78	0.61	0.83	0.89	5.95
	H	0.17	0.28	0.28	0.33	0.50	0.50	0.61	0.67	0.78	0.83	4.95
	F	0.17	0.22	0.28	0.33	0.50	0.50	0.61	0.67	0.72	0.83	4.83
	A	0.22	0.22	0.33	0.22	0.39	0.39	0.50	0.72	0.72	0.67	4.38
	C	0.17	0.28	0.39	0.39	0.33	0.33	0.28	0.50	0.61	0.67	3.95
	J	0.11	0.06	0.11	0.17	0.22	0.28	0.28	0.39	0.50	0.44	2.56
	B	0.17	0.06	0.11	0.11	0.17	0.17	0.33	0.33	0.56	0.50	2.51

Law of Comparative Judgment

Prior to describing the actual operations involved in arriving at an equal interval measurement scale of representativeness of the artistic sketches, it is necessary to consider the logic underlying these procedures. The basic principle, known as the *Law of Comparative Judgment,* was first elaborated by Thurstone (1927).

Essentially, it is assumed that the psychological scale separation between two stimuli is a normal function of the observed proportion of times that one of the stimuli is preferred to the other. Since a normal distribution of discriminative reactions is assumed to underlie the observed proportions, an equal interval scale can be achieved by converting the proportions to normal–deviates.

Just as was done in determining thresholds by the *Method of Constant Stimuli,* what must be done to provide standard units of distance between pairs of stimuli is to transform the proportions of area under the normal curve (assumed to be reflected in the observed proportions of preference) into z distances on the abscissa of the normal curve.

This logic is actually nothing more than an extension of the thinking of the classical psychophysicist, as exemplified in Fechner's famous psychophysical law. In order to derive his equation, $R = c \log S$, from Weber's constant fraction, Fechner had to assume that the size of the JND (just noticeable difference) is the same at all points on the stimulus scale. But this assumption was easy to make, simply because thO is always defined in terms of the proportion of times that a stiis is judged greater (louder, brighter, etc.) than another. If one of peas is judged to be heavier than another 0.75 of the time, and if on two golf balls is judged to be heavier than another 0.75 of the time, dis- tance in psychological "heaviness" is the same in both compari In the present example, the psychological representativeness-dist be- tween two sketches, one of which is judged to be more repreative 0.83 of the time (e.g., sketch E versus sketch H) is equal to thecho- logical distance between two other sketches, one of which is jd to be more representative 0.83 of the time (e.g., sketch D versus sh J). This is the well-known Fullerton-Cattell principle that "equaften noticed differences are equal."

Weighting the z Values

However, the normal distributions involved in these two differcom- parisons may involve different scale units, since the discriminatreac- tions in the two different cases may have different standard dtions (after all, $z = x/\sigma$). Thus, the z values obtained by convertinppor- tions into normal–deviates may have to be differentially weigl (for the differences in standard deviations and differences in intercotions of responses to the stimuli) for the different pairs of stimuli. Thethod to be presented below, however, is based on the simplifying asption that differential weighting is unnecessary in the present situat (this assumption is known as Thurstone's Case V). The normal–devs cor- responding to the proportions given in Table 4.3 are presented Table 4.4.

The extreme right-hand column of Table 4.4 presents the meof the z values for each of the ten sketches. These values locate thetches on an equal interval scale of representativeness, based on the asptions previously discussed. The reason that the entire set of z's for easketch is used to determine its scale value is simply that the averapf the z distances between the sketch in a given row and all the otheretches gives the average z location of the particular sketch. This stemsm the fact that the distance between a pair of sketches may be estied not only from the proportion obtained when they are compared dily but from the differences between the obtained proportions wheney are each compared to a third stimulus. Adding up all of the z disces for each stimulus and obtaining the mean as an estimate of its sc value has the effect of locating each of the stimuli in relation to of the others. In the present case, since we have no information regng the true zero of the scale of representativeness, it would be sufficien add a

Normal–deviate (z) matrix for paired comparison scaling of representativeness of ten artistic sketches, listed in the same order as in Table 4.3.

	Sketches											
	E	G	I	D	H	F	A	C	J	B	Σ	$\bar{X}$
E	0	0.44	0.58	0.77	0.95	0.95	0.77	0.95	1.23	0.95	7.59	0.76
G	–0.44	0	0	0.28	0.58	0.77	0.77	0.58	1.55	1.55	5.64	0.56
I	–0.58	0	0	0.28	0.58	0.58	0.44	0.28	1.23	1.23	4.04	0.40
D	–0.77	–0.28	–0.28	0	0.44	0.44	0.77	0.28	0.95	1.23	2.78	0.28
H	–0.95	–0.58	–0.58	–0.44	0	0	0.28	0.44	0.77	0.95	–0.11	–0.01
F	–0.95	–0.77	–0.44	–0.44	0	0	0.28	0.44	0.58	0.95	–0.49	–0.05
A	–0.77	–0.77	–0.28	–0.77	–0.28	–0.28	0	0.58	0.58	0.44	–1.71	–0.17
C	–0.95	–0.58	–1.23	–0.28	–0.44	–0.44	–0.58	0	0.28	0.44	–2.83	–0.28
J	–1.23	–1.55	–1.23	–0.95	–0.77	–0.58	–0.58	–0.28	0	–0.15	–7.32	–0.73
B	–0.95	–1.55	–1.23	–1.23	–0.95	–0.95	–0.44	–0.44	0.15	0	–7.59	–0.76

constant of 0.76 to all of the mean z values, which has the effect of locating the *least* representative sketch (sketch B) at the bottom of the scale with a value of zero (and all the other values are then positive).

In the present example, there would be no feasible way in which the actual zero on the representativeness scale could be located. It would be possible to add a sketch to the judgment task, which was made of some other subject matter, or one that was made entirely non-representationally without explicit subject matter. But the choice of a completely non-representative sketch would be somewhat arbitrary. A better method in this situation might be to have the subjects compare the sketches with an additional one that was traced from the actual photograph (and, thus, could be taken to represent *perfect* representativeness). Thus, while no true zero would be available, perfect representativeness, expressed as *unity*, would be. The scale values could then arbitrarily be transformed linearly after the traced sketch was arbitrarily assigned a scale value of 1.0, so that they fell between 1.0 and 0.

Internal Consistency of Paired Comparison Scale Values

The average z values are taken to be an equal interval scale of representativeness of the sketches. It is assumed that the various comparative judgments do not reflect different underlying normal distributions. This assumption may be approximately empirically testing. Given the resulting set of scale values, a new matrix of *expected* z values for pairs of stimuli can be generated from them by subtracting the obtained average z of one stimulus from the average z of each other stimulus (i.e., by obtaining scale separation values). These *expected* z values are then translated back into *expected* proportions for comparison with the original observed proportions.

The question being asked here is whether the proportions expected on the basis of the resulting average scale values deviate significantly from those originally obtained. Table 4.5 presents the scale separations (obtained by subtraction, using the values in the extreme right-hand column of Table 4.4) and Table 4.6 gives the expected proportions generated by these z values. Since both of these tables are diagonally symmetrical, only one half of the values are necessary (omitting the principal diagonal, for which no actual data were collected).

Test of Significance

Comparison of the *expected* proportions in Table 4.6 with the corresponding *observed* proportions in Table 4.3 indicates that a substantial

TABLE 4.5

Expected z values (scale separations) obtained by subtraction of average scale values of pairs of stimuli (extreme right-hand column of Table 4.4).

	Sketches								
Sketches	G	I	D	H	F	A	C	J	B
E	0.20	0.36	0.48	0.77	0.81	0.93	1.04	1.49	1.52
G		0.16	0.28	0.57	0.61	0.73	0.84	1.29	1.32
I			0.12	0.41	0.45	0.57	0.68	1.13	1.16
D				0.29	0.32	0.45	0.56	1.01	1.04
H					0.03	0.16	0.27	0.72	0.75
F						0.13	0.24	0.69	0.72
A							0.11	0.56	0.59
C								0.45	0.48
J									0.03

TABLE 4.6

Expected proportions generated from z values in Table 4.5.

	Sketches								
Sketches	G	I	D	H	F	A	C	J	B
E	0.58	0.64	0.68	0.78	0.79	0.82	0.85	0.93	0.94
G		0.56	0.61	0.72	0.73	0.77	0.80	0.90	0.91
I			0.55	0.66	0.67	0.72	0.75	0.87	0.88
D				0.61	0.63	0.67	0.71	0.84	0.85
H					0.51	0.56	0.61	0.76	0.77
F						0.55	0.59	0.75	0.76
A							0.54	0.71	0.72
C								0.67	0.68
J									0.51

degree of internal consistency was present in the original data. A Chi Square test of the significance of the difference between the expected and observed proportions (Mosteller, 1951), in fact, indicates that the discrepancies could have occurred by chance as often as 0.95 or more of the time (Chi Square = 17.94, *df* = 36). Under these conditions, the assumption made in determining the scale values from the original proportions are probably quite justified.

The logic of Thurstone's scaling method has been extended to situations in which stimuli are placed in successive categories (the *law of categorical judgment*), as well as to cases in which stimuli are actually ranked as a group rather than judged in pairs. In the latter situation, the ranked data are converted to "pseudo-paired comparisons" prior to the z transformation (each item is given a proportion depending upon the number of times its rank is higher than that of each other item). In addition, as is noted below, scaling of arbitrarily measured responses by Thurstone's method has been used with some success. From the present point of view, application of Thurstone's method to response-scaling has the potential to add significantly to the level of quantification and rationalization appropriate to the importance of the subject matter of scientific psychology.

Scaling from Ratio Judgments

Perhaps the most sophisticated attempt to develop fundamental measurements of psychological reactions to stimuli is to be found in the work of Stevens (1961) and his colleagues at the Harvard Psychoacoustics Laboratory. Their *sone* and *mel* scales of loudness and pitch of tones, respectively, have become standard devices for the description of auditory stimuli in psychological experiments, as well as contributing significantly to our understanding of basic perceptual processes.

The fundamental procedural devices employed by Stevens, involving either multiplication or fractionation of stimulus attributes, assumes that the subject is able to form ratios of sensation directly. Although this is an a priori attribution of ability to the subject and, as such, raises serious questions regarding whether the subject can actually perform the assigned task, the voluminous data indicating that reliable and internally consistent psychophysical scales can be obtained in this way should satisfy even the most skeptical critic.

As was noted in Chapter 3, the actual measurement operations (e.g., producing a tone half or twice as loud as a standard tone, or producing a tone that bisects the loudness difference between two standard tones) are not, themselves, direct empirical implications of Stevens' theory of loudness perception. Yet, the methods he has used become examples of fundamental measurement to the extent that fractionating or multiplying experiences of loudness results in internally consistent measurement scales, since no other fundamentally measured dimensions are involved and the arbitrariness of the assumed ability lends itself to empirical check (e.g., the demonstrated repeatability of Stevens' obtained power functions).[5]

To illustrate the similarity between Stevens' methods and the fundamental measurement of temperature described in Chapter 3, an example of fractionation will be considered. A subject is instructed that his task is to produce a tone whose loudness exactly bisects the subjective loudness distance between two standard tones, A and E. He does this by manually manipulating a continuously variable control, occasionally listening to the variable tone (as well as "reminding" himself of the two standard tones by listening to them periodically). The tone he produces will be referred to as tone C. The procedure is then repeated using tones A and C as standards, so that tone B is produced (bisecting the loudness interval between A and C). Similarly, tone D is produced so that it bisects the loudness interval between tones C and E. The test of internal consistency is achieved by having the subject bisect the interval between tones B and D to produce a new tone, one that should have the same subjective intensity as tone C.

A ratio scale of loudness is produced by this method (since an absolute zero is associated with the absolute threshold and, since the subject's division of the attribute into equal steps assumes the ability to make and perceive ratios). The actual units of the scale (sones) are arbitrarily anchored to a physical standard (i.e., one sone is the loudness of a 1,000 Hz tone at 40 decibels above threshold). In the temperature example, where bisection is achieved by mixing of equal amounts of water at two temperatures, it is achieved in the loudness case by assuming that the subject can produce the appropriate bisecting tones. The test of internal consistency establishes whether or not he actually can.

SCALING OF RESPONSES

Several references have been made in the preceding pages to the fact that the vast majority of behavioristically oriented studies in the pub-

[5] A number of other methods of direct production of ratio scales have also been used in the scaling of stimuli. One of these, the *constant sum method*, is of interest because it may be used with qualitatively different stimuli as easily as with quantitative stimulus properties. Since comparisons between the constant sum and fractionation methods (Guilford, 1954) indicate that similar psychophysical equations are produced by both when the same stimuli are used, it may be that a variety of qualitative stimulus dimensions can be psychologically scaled to have ratio scale properties, using the constant sum technique. The subject's task in the constant sum method is to divide a fixed number of points (e.g., 100) between two or more stimuli in such a way that the division reflects the ratio of their subjective values. The quantitative details of this procedure are given by Comrey (1950).

lished literature have used more or less arbitrarily measured dependent variables, more often than not employing physically defined dimensions such as response *latency, force, amplitude,* etc. A great deal of objectivity and reliability is gained in this way, since the measurement scales of physical science are very familiar and readily adaptable for use by even inexperienced researchers. Results expressed in milliseconds or pounds per square inch have the semblance of real science. The fact remains, however, that psychologists are more often interested (or should be) in the *psychological* properties that are arbitrarily indexed by these physical quantities (with exceptions as noted previously) than they are in the physical measures themselves.

The problem is not one that invites easy solution; nor is the simple criticism, unaccompanied by feasible suggestions for improvement, likely to be either well received or responded to seriously. Clearly, its essence is that of basic scientific measurement; yet, those psychologists who have been most closely identified with psychometrics in recent years have not come from the ranks of the hypothesis-testing behavioristic experimenters as often as they have from the correlational and psychophysical scaling areas.

In the context of a simplified example, the question may be asked whether an experimental subject, such as a rat, which performs an approach response (say, running down an alley for a food reward) in two seconds has twice as much (or how many times as much?) of the "tendency" to perform this response as another rat, which takes four seconds to reach the goal box? In the ratio scale of the physically measured units of time, four seconds *is* twice the amount of time as two seconds. Yet, we have every reason to believe that the strength of the "tendency" to perform the response is not a linear function of the time measure. Nor is this abstract potential for responding (*reaction potential,* in the older language of Hull's theory) a simple linear function of grams of pressure, ohms of resistance change, number of errors, etc., in those situations in which these other ratio scale quantities are recorded. As Hull et al. (1947) have observed, "All these presumably possess a monotonic relationship to reaction potential, and all are evidently increasing functions of it except reaction latency and number of errors which obviously have a decreasing or inverse functional relationship. But, ${}_{S}E_{R}$ (reaction potential) itself clearly is not necessarily a certain number of grams, millimeters, cubic centimeters, seconds, or errors" (1947, p. 238).

At the level of simple hypothesis-testing research, this nonlinear mismatch between the measurement scale we actually employ and the attribute that commands our more fervid interest may produce few serious misunderstandings. By and large, we should still be able to detect significant effects of potent independent variables, as, in fact, we so often have. But, at the more austere level to which we aspire, the level of estimation experiments and experiments purporting to evaluate the specific quantitative details of elaborate theories, the use of arbitrary indices borrowed from physical science cannot but significantly mislead and confuse us. How, for example, could we ever expect to compare learning or habituation curves based on different physical measures?

Scale of Reaction Potential

An interesting and pioneering attempt to use a standard psychological scaling technique to develop an equal interval scale of the strength of reaction potential from units of response time was made by Hull et al. (1947). Beginning with response times of a group of 59 rats, measured from the moment of presentation of a small manipulandum until it was moved 1/16 inch to the left, each rat's response time on a given trial was compared with its response times on other trials, to generate a matrix of frequencies of occasions on which each response was faster than each of the others. These frequencies were summed for all 59 rats and converted to proportions.

The data for nine consecutive trials at the beginning of a learning situation (for illustrative purposes) were, thus, represented as a proportion matrix, where each proportion in a cell reflected the relative frequency with which the rats tended to respond faster on the trial in that row compared with the trial in that column. These proportions were then converted to normal–deviates, in the fashion of Thurstone's method of scaling from comparative judgments. It should be clear that the trials were being treated much in the fashion in which stimuli are treated in Thurstone's method.

Since reaction latency can be shown to vary in random fashion from trial to trial when habit and drive factors remain constant, it seemed reasonable to Hull et al. to assume the required normal distribution of reaction potentials, upon the basis of which the transformation from proportions to normal–deviates is made.

Arbitrarily assigning a value of zero reaction potential to the first trial (since no learning is assumed to have occurred at this point), the resulting adjusted scale values of reaction potential are presented in Figure 4.6 (it should be pointed out that, while Hull et al. used Thurstone's Case III to estimate the standard deviations for each trial and adjusted the scale values accordingly, the present simplified illustration proceeds on Thurstone's Case V assumption that such adjustment is unnecessary). As Hull et al. point out, the scale values of ${}_sE_r$ in Figure 4.6 reflect an operational definition of the concept of reaction potential.

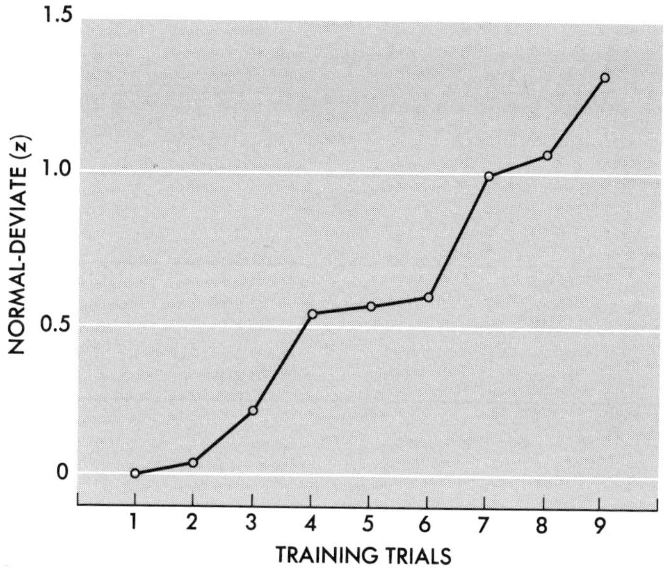

Figure 4.6 Relationship between ${}_sE_r$ and number of training trials, based upon paired comparison scaling of rats' response times. Data from Hull et al. (1947).

Comparison of Learning Curves

The potential usefulness of this application of paired comparison scaling (beyond what Hull et al. have demonstrated) may be illustrated if we *compare* the growth of reaction potential over conditioning trials when the scale values are based on the amplitude of the galvanic skin response in humans with the scaled values based on response times of Hull's rats.

Using the change in the logarithm of conductance as an index of amplitude, a proportion matrix based on five classical conditioning trials was constructed, as is shown in Table 4.7. Reading from left to right in each row in this table, the proportions indicate the relative frequency with which the galvanic skin response of the trial in that row was larger than the galvanic skin response of the trial in each column. These proportions are converted to normal–deviates, as is indicated in Table 4.8. Also shown in the two extreme right hand columns of Table 4.8 are the sums and means of the scale values. Assigning an arbitrary value of zero to the first trial (since no pairing of CS and UCS had yet occurred), the resulting adjusted scale values are plotted in Figure 4.7.

In comparing the curve obtained in this way with the one shown in Figure 4.6, it should be recognized that the overall *levels* of the two may

TABLE 4.7

Proportion matrix based on amplitudes of the galvanic skin response of 46 human subjects in five trials of classical conditioning.

		Trials				
		1	2	3	4	5
Trials	1	0.50	0.38	0.33	0.15	0.20
	2	0.62	0.50	0.36	0.42	0.40
	3	0.67	0.64	0.50	0.48	0.51
	4	0.85	0.58	0.52	0.50	0.52
	5	0.80	0.60	0.49	0.48	0.50

TABLE 4.8

Normal–deviate (z) matrix based on proportion matrix of Table 4.7, assuming Thurstone's Case V.

		Trials						
		1	2	3	4	5	Σ	$\bar{X}$
Trials	1	0	–0.31	–0.44	–1.04	–0.84	–2.63	–0.53
	2	0.31	0	–0.36	–0.20	–0.25	–0.50	–0.10
	3	0.44	0.36	0	–0.05	0.03	0.78	0.16
	4	1.04	0.20	0.05	0	0.05	1.34	0.27
	5	0.84	0.25	–0.03	–0.05	0	1.01	0.20

simply reflect differences in the units of the two scales, since even under the simplifying assumptions of the Case V scaling of the galvanic skin responses there is no reason to assume that the normal distributions of these measures have the same standard deviation as those of the rats' response times. However, the curves' *shapes* (for the first five trials) are quite similar, suggesting that they may, in fact, reflect the same kind of underlying habit formation process (the slight decline in the fifth trial of the humans' galvanic skin response scale values may be due to the development of a secondary mechanism that is peculiar to the classical conditioning procedure).

The potential *comparative* usefulness of this method of paired-comparison response scaling would appear to be aptly demonstrated in the fore-

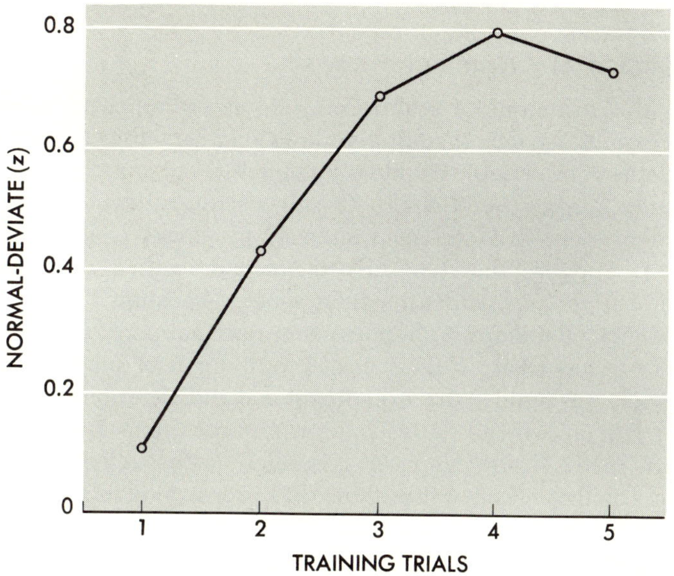

Figure 4.7 Relationship between reaction potential and number of training trials, based upon paired comparison scaling of galvanic skin response amplitudes of 46 human subjects (unpublished).

going examples.[6] However, the use there of *trials* as the basis of comparison would seem to prevent an immediate extension of the method to the more general problem of determining the scale positions of different responses (*either* among different subjects *or* among occasions within the same subject) relative to one another. In the case of scaling of *stimuli,* based on responses to them, the *stimuli* find psychological definition thereby. *Responses* cannot be scaled by either Thurstone's or Stevens' methods without reference to either trials (occasions) or subjects. To be accurate, in the Hull et al. example, *trials* were being scaled in terms of responses. This limitation notwithstanding, *subjects* can be scaled in terms of their responses. Thus, over a given set of trials it may be said that subject A was twice as high as subject B, etc. (if a ratio scale could be obtained), or that subject A was twice as much better than subject B as he was better than subject C (if an interval scale was obtained). Scaling of *subjects* would, of course, require repetition of measurement across occasions (or trials), so that a proportion of occasions of superiority of one subject over another could be obtained (just as scaling of *trials* required repetition across subjects).

[6] Additional examples of comparative scaling of responses are found in Kimmel, Boice, and Leckart (1969 a,b).

Determination of Quantitative Laws

Using this technique of scaling subjects according to their responses across occasions, specific quantitative laws relating subjects' responses to some independent variable could be obtained. For example, the terminal performance (over several trials) of subjects run under different levels of an independent variable (e.g., intensity of shock) could be obtained as follows. Suppose that ten subjects are run with a weak shock (2 mamps.), ten are run with a medium shock (4 mamps.), and ten with a strong shock (6 mamps.). Suppose that terminal performance is measured on forty test trials. Taking each possible pair of subjects (ignoring which shock was used), the experimenter could determine the proportion of *trials* on which A's response was larger than B's. A 30×30 proportion matrix could, thus, be generated, with subjects forming the rows and columns. Converting these 900 proportions into normal–deviates would be done as previously.

Because of the difference in shock intensity among the three groups, it would probably be necessary to employ Thurstone's Case III (i.e., estimates of standard deviations for each subject would be needed as a basis for weighting the z values). The final average scale values of the three groups would provide a basis for a quantitative law relating intensity of the shock and terminal reactivity. Analogous extension of this method could be made to all psychological phenomena with known parameters (e.g., classical and instrumental conditioning, rote learning, habituation, animal motivation, etc.).

SUMMARY

In *psychophysical* research, the dependent variables are expressed as *stimulus quantities*, with the subject's response functioning as an indicator of either the presence or absence of the stimulus or the similarity or difference between stimuli. The major traditional methods of psychophysics are the *method of limits*, the *method of constant stimuli*, and the *method of average error*. These methods are used to determine *absolute* and *difference thresholds* and the *point of subjective equality*. The method of constant stimuli was developed to overcome the errors of *expectation* and *habituation* which attenuated the effectiveness of the method of limits.

A basic assumption underlying most psychophysical research, the *phi–gamma* assumption, is that the relationship between response and stimuli in the region of the threshold follows the *normal curve*. On the basis of this assumption, a *psychophysical function* relating responses to stimulus variations may be fitted using the *method of least squares.* From the fitted psychophysical function, thresholds and other dependent measures are obtained.

Psychological reactions to stimuli are used to develop *equal interval* or *ratio scales* of stimulus properties. The method of *paired comparisons,* developed by Thurstone, and the *ratio scaling* methods, developed by Stevens, are most significant in this area. In addition to scaling the psychological properties of stimuli, these methods may be used to develop *response-scales* for comparison of results from studies using different arbitrary response measures or those dealing with hypothetical response processes, although these applications are relatively recent and unexploited.

5

Choosing the Independent Variables

PROBLEM ANALYSIS

Problem analysis begins at the moment of conception of the explanatory generalization whose empirical validity ultimately must be tested experimentally. As soon as the idea first comes to mind that a particular set of observations may be explained in a certain way, the explanation should imply the manner of its empirical evaluation. The logical process involved is both synthetic and analytical, as if the inductive fusion and deductive splitting of the explanatory atom each depended for its existence upon the other.

Experimental testing of a particular theoretical conceptualization, however, may be achieved in more than a single way. The more loosely structured the idea, the more different ways there are for assessing it empirically. The more highly structured the idea, the narrower the range of different experimental outcomes it implies, although their specific details may be more fully spelled out. Even when an explanatory generalization is part of an organized system of related concepts (and, thus, is highly formalized), the imaginative experimenter will find that its implications generate many different experiments or many different predictions within the confines of a single, complex experiment.

Among the most important elements of problem analysis is the matter of defining the dependent variable to be observed. As was seen in Chapters 3 and 4, a substantial part of problem analysis concerns the way in which the dependent behavior measurement procedures are specified operationally. It was convenient to deal with this aspect of problem analysis separately, as was done in Chapters 3 and 4. In the present chapter, problem analysis will be considered in relation to the broad issue of *choosing* the independent and dependent variables. To do this, examples of particular experiments will be examined which illustrate significant aspects of problem analysis.

EXPERIMENT ON FORGETTING

When a person learns a list of words or nonsense syllables, or when he memorizes a poem, his ability to recall the learned material may suffer substantial impairment over time without practice. Ever since the classical work of Ebbinghaus (1885) on memory, the form of this temporal decay function has been well known and its explanation the topic of extensive study. In the early days of this investigation, the idea that the passage of time, itself, was somehow responsible for the memory loss was thought to be credible. In fact, the reduction in retention over time was said to describe the "law of disuse," as if that which was established by the original learning gradually weakened merely because it was *not* rehearsed or practiced over some time period.

There were many (e.g., McGeoch and Irion, 1952), however, who felt that forgetting was more appropriately attributable to an active interference process, rather than to passive diminution over time. Time played the role, in this view, solely of providing a framework within which new events could occur which would have the effect of impairing the original learning.

To evaluate this interference theory of forgetting, a way was needed in which to hold the passage of time constant while deliberately arranging for the degree of active experience following the original learning to vary. Analysis of the problem quickly reduces it to the question of how to vary the degree of experience following learning, while not inadvertently permitting other potentially influential factors to vary simultaneously. Although there are undoubtedly many ways in which this goal could be achieved, one of the most imaginative was the method employed by Jenkins and Dallenbach (1924), in their famous experiment on retention. Two subjects learned the same material to equivalent criteria of mastery. Both subjects were tested for retention at various hours after the original learning session. However, while one of these subjects passed this time going about the routine business of daily life as a college student, the other subject, who would conventionally be referred to as the *experimental subject,* spent the time in bed, sleeping.

As can be seen in Figure 5.1, the experimental subject showed substantially greater retention (less forgetting) than did the former (*control*)

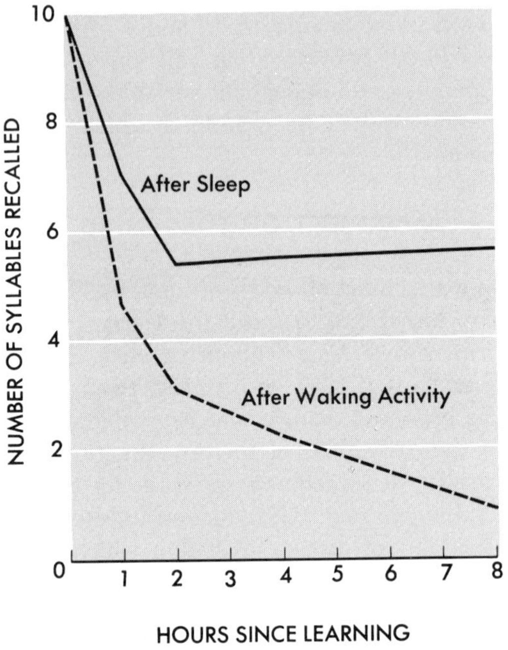

Figure 5.1 Forgetting after sleep and waking. From data in Jenkins and Dallenbach (1924).

subject, supporting the notion that an active interference process is responsible for memory loss, rather than the mere passage of time since learning.

Obviously, this experiment is not without its methodological imperfections (When does the sleeping take place in relation to previous sleeping? Perhaps sleep, itself, involves some active, consolidating process, which prevents the otherwise passive loss over time. Are the two subjects typical of some larger group? etc.). But it is useful to us in exemplifying the way in which the suggestion of a manipulable independent variable appears to emerge naturally from the very conceptual substance of the theory to be tested. The theory in this case states that forgetting is due to an active interference of post-learning events and experiences with previous learning.

The experimental manipulation of the amount of such interference is immediately implied, although, as has been observed, the use of *sleep* to represent minimal interference was a unique manifestation of the experimenters' creativity. To be sure, the theory as stated implies nothing about which *types* of events would be more likely to produce interference and which would be less likely to do so. That is, nothing is suggested regarding the manner of operation of the hypothesized interference. From the present point of view, this is a deficiency in experimental planning, an example of inadequate problem analysis. It may be assumed that an experimenter who supposes that *something* associated with ordinary daily activities acts retroactively to interfere with the retention of previously learned material (as compared to what happens during sleep) would be able to make explicit also just what that something might be and how it might work. If he can say *why* he expects sleep to reduce the amount of interference, he may be able to improve his experiment considerably.

In the present case, for example, the performance of the subject's routine activities might be expected to reduce retention of previously learned material to the extent to which these activities are somehow related to the learned behavior. If the learning is thought of as involving associations between stimuli and responses, it may be that the stimuli that occur during ordinary daily living are similar to those used in the experiment and produce interference because they are associated with responses different from those learned in the experimental situation. These new responses might interfere with the previously learned ones because they are mutually incompatible. On the other hand, rather than the *similarity* of daily and experimental stimuli, it may be that the sheer *number* of the former is what is significant in forgetting. This line of thinking suggests that there might be a limit to the amount of material capable of being stored at one time and that the newer material might gradually "push" the older material out of the memory storage.

Two different lines of reasoning are implied, the first suggesting manipulation of the *degree of similarity* of the learned and interpolated material and the second suggesting manipulation of the *amount* of interpolated material. Implementation of either or both of these additional independent variations, as an augmentation of the original Jenkins and Dallenbach experiment, might have clarified some of the follow-up questions raised by its results.

EXPERIMENT ON MOTIVATIONAL FACTORS IN PERCEPTION

About 25 years ago a group of American psychologists initiated a theoretical and experimental program centered around the then novel idea that perception involves a dynamic interaction between the perceiver and the perceived, rather than being determined solely either by physically definable characteristics of stimuli or by physiologically defined characteristics of receptor–analyzer systems. In addition to these *autochthonous* factors in perception (i.e., physical and physiological), a broad domain of *behavioral* factors (e.g., learning, motivation, etc.) were said to play an important role in what is perceived. This *transactionalist* approach (Postman, 1953), also sometimes referred to informally as the "new look in perception," generated many ingenious and controversial studies on the influence of associative and motivational variables on perception. To be sure, not all of these experiments succeeded in overcoming the rather formidable methodological obstacles inherent in studies of covert hypothetical processes such as perception. For the purpose of using one of these studies to elucidate the way in which problem analysis may result in extension of the empirical scope of an experiment, however, we will ignore many of the serious methodological problems in this type of research.

The initial question that must be raised by the experimenter, as he begins to develop an experimental plan for testing the empirical soundness of an explanatory generalization, is, "What are the potentially observable implications of this explanation?" At first glance, it might seem that the posing of a question of this type would hardly be necessary for an experimenter whose theorizing quickly generates ideas for empirical testing. But, even though it may be true that such ideas are present at least implicitly from the very beginning of the inductive process, only by *explicitly* asking and *fully* answering the question can the scientist be reasonably certain that the experiment he ends up doing is really the most effective one for his purposes. Too many experiments are capable of being described as "only scratching the surface," when, with more thorough analysis of the implications of the explanations they attempted to test, they could have been greatly improved without an impractical amount of elaboration.

In the case of the experiment on forgetting, problem analysis was said to be incomplete in that the explanatory conceptualization was not elaborated sufficiently to ensure that positive results would enable us to know *why* interpolated activity results in memory loss. This type of deficiency may occur in even more exaggerated form, for example, when the experimenter denies that *any* deductive reasoning played a role in his pre-experimental behavior (Skinner, 1956). On the premise that the goal of experimentation is "control" of behavior (or prediction), rather than theoretical explanation, the argument is sometimes made that it doesn't matter *how* or *why* the independent variable has the effect observed, it matters only *that* it has the effect. A kind of "know-nothingism" with respect to explanation is the inevitable result of this posture, unfortunately, one of the consequences of which is a proliferation of single-minded experiments and only sketchily related findings. Another consequence, it may be assumed, is a failure to define the conditions for experimentation so that the technology of behavioral control can actually develop efficiently. As Skinner has himself remarked elsewhere, ". . . the power of any technology depends upon an *understanding* of its basic processes" (1968, p. 704, emphasis added).

Assume the Explanation Is Correct

A useful way to approach answering the question of potentially testable implications of a proposed theoretical explanation is for the experimenter to start by assuming that his explanation is actually completely correct! He is then in the position of asking, "Given that this *is* the way in which the process really works, what kinds of outcomes should we expect under various experimental conditions?" "*Which* manipulable independent variables would be likely to influence *which* dependent measures and in *what* ways?" The more exhaustively this line of questioning can be pursued, the more likely it is that the experimenter will succeed in placing his theoretical concepts under sufficiently sensitive empirical scrutiny. Of course, each additional testable implication of the theory that is built into an experiment adds to the probability of discovering that the explanation is incorrect, since each separate test provides an additional opportunity for observed disagreement between predictions from the theory and actual observed behavior. But, this is what experimental research is all about anyway and the dedicated investigator quite willingly exposes his explanatory ideas to the danger of falsification by experimental facts.

Perception of the Size of Coins

Bruner and Goodman (1947) conducted a much-discussed experiment on the role of motivational factors in the perception of the size of objects. Without presuming to describe the actual sequence of thought that led to the final procedures used (since these are not known to us), it is instructive to examine what *might* have been the conceptual progression of these authors.

Beginning with the basic idea that what is seen is the result of a dynamic "transaction" between the perceiver and the physical stimulus, it may be deduced that valued objects will tend to be enhanced perceptually in comparison with nonvalued objects. This enhancement, however, could be manifested in a number of different ways, only one of which, perceived size, was actually examined in this particular experiment. For example, subjects might report that the perceived duration of presentation of valued objects was greater than that of nonvalued objects, when both types of stimulus actually were presented for equal time periods. Or, it might be expected that subjects would overestimate the heaviness or brightness of valued objects.

Another possible manifestation of perceptual enhancement of valued versus nonvalued objects might be that subjects would choose to look longer at the former as compared with the latter, in a "looking-time" experimental task. It is instructive to note that the last suggested method would lead to a completely different kind of dependent variable than any of the others, since the size, heaviness, and brightness approaches all suggest a psychophysical dependent variable, with stimulus quantities being used to describe a standardized, dichotomous (digital) response, while the looking-time method uses standardized stimuli and measures quantitative (analog) variations in responses. As was shown in Chapter 3, this difference in method of defining the dependent variable is the basis for a number of other differences in both methodology and interpretation of results.

The actual method chosen by Bruner and Goodman was to have their subjects match the size of a variable spot of light on a screen with that of an object either held in the hand or shown at close range. Coins were used as valued objects and pieces of cardboard of the same size as coins were used as nonvalued objects. Each type of object was presented to the subject several times, using the psychophysical method of average error. The principal implication of their theoretical reasoning called simply for an observed difference in the subjects' average adjustment of the size of the variable stimulus (the spot of light on the screen) when it was being matched in size with a coin as compared with when it was being matched with a cardboard disc. That is, if the perceived size of *valued* objects is enhanced, Bruner and Goodman reasoned that the subjects should adjust the variable spot of light to a larger size when it was being matched to a coin than when matched to a cardboard disc. Given the particular methodology employed, this difference would seem to be clearly implied by the theory from which Bruner and Goodman began.

Dimensional Analysis

With the consequence of doing a much better experiment, Bruner and Goodman pursued the analysis of their problem well beyond this first point of possible testing. It would have been entirely possible to compare average settings of the variable stimulus under the coin and cardboard disc conditions, without further elaboration of additional implications of the theory. But the theory implies not only that valued objects should be seen as larger than nonvalued objects, but that this difference should be greater for *more* valuable coins than for *less* valuable coins. This particular result of problem analysis, in which a quantitative independent variable may be used to extend the range of testing of another independent variable (whether quantitatively of qualitatively defined), may be labeled *dimensional analysis.*

Without adding unnecessary complexity to the subjects' task (in fact, probably succeeding in making it far more interesting), Bruner and Goodman used coins ranging in value from one cent to fifty cents, as well as cardboard discs appropriately equal in size to each of these. With this extension in their experimental procedure (i.e., using coins of several values rather than simply using a coin of a single value), their theory required, first, that coins be generally perceived as larger in size than the cardboard discs and, in addition, that the degree of this enhancement increase as the value of the coin increased. *More* valued objects should have their perceived size enhanced *more* than less valued objects, albeit that all will manifest some enhancement. As will be seen below, the increased jeopardy into which their theory was placed by this embellishment of their experimental procedure, had the unexpected beneficial effect of revealing the possible influence of a complicating factor, the subjects' experience with coins.

Because of the need to employ stimuli whose value for the subjects could be safely assumed, and because it was necessary that the task be sufficiently interesting for the subjects to sustain cooperative performance, Bruner and Goodman decided to use elementary school children in the experiment. It was expected that the children would be more likely than adults to be sensitive to the relatively small monetary values involved, especially considering that even pennies were used in the study. Just below the surface of this reasoning is the implication that still another independent variable might be appropriate for use, namely, the difference in the extent to which different subjects *need* money (and, therefore, value it). To employ this type of independent variable, obviously, would require *selection* of subjects on the basis of some criterion reasonably assumed to be highly correlated with the need for money.

The relationship between variations in the value of coins, on the one hand, and variations in subjects' need, on the other, is somewhat analogous to that between variations in duration of food deprivation and in amount of food reward in instrumental learning studies. In both cases, a relationship between the internal and external motivational variables underlies both types of independent variation. However, in the deprivation–food situation, both variables are directly manipulable by the experimenter, while in the value–need situation, value is manipulable and need is not.

For reasons already presented in Chapter 2, only the actually manipulable value variable would appear to be eligible to enter into causal arguments, although in this case its dependence upon the need variable muddies the waters somewhat (Can value be defined completely independently of need?). In any case, the intimate bond between value and need in this situation (the one manipulable and the other only selectable) may serve to illuminate one aspect of the problem involved in our previous attempt to distinguish between experiments and quasi-experiments.

Bruner and Goodman decided to add this subject variable as a formal part of their experiment by using children from two different schools, one located in an economically depressed area and one in an upper middle class area. It was quite reasonable for them to assume that the children in one school would differ, on the average, from those in the other in the degree to which small coins were valued by them, since it was likely that the poorer children's families exhibited greater daily concern over monetary matters (even pennies) than would occur in the wealthier families.

What began as a straightforward prediction that their subjects would perceive coins as larger than nonvalued objects (requiring only one kind of subject and one kind of coin), because of effective problem analysis, became a complex experiment with three independent variables, object-value, amount of object-value, and type of subject. In the format of this experiment, the theory now required not only that coins be seen as larger than discs, that this difference in perceived size increase as the value of the coin increased, but that the poor children show the effect more than the rich children.

Confounding by Subject Variables

The assumption that children from economically well-off families would differ from those from poorer families in the extent to which coins represent valuable objects was, as has been noted above, not at all unreasonable. In fact, Bruner and Goodman's finding that the poorer children *did* tend to overestimate the size of coins somewhat more than the richer children is in accord with this assumption. However, the choice of a subject variable in this instance illustrates one of the ubiquitous dangers inherent in this procedure, since it was likely that the poorer children differed from the richer children in other aspects of their experience with coins, particularly in their familiarity with them. Confounding by subject variables other than the one intended to vary by the experimenter is the most common pitfall associated with the use of selected rather than manipulated independent variables.

The results of this study are shown in Figure 5.2. That familiarity with the coins may have been associated with the difference between the groups (along with the intended difference in need for coins) is supported by the finding in this study that, while all of the children tended to overestimate all of the coins in relation to the discs, and while this effect tended to increase with increases in the value of the coins (both of these as predicted by the theory), an inversion in the latter tendency was found between the twenty-five and fifty cent pieces. The difference between the perceived size of the twenty-five cent piece and its cardboard disc was greater than the comparable difference between the perceived size of the

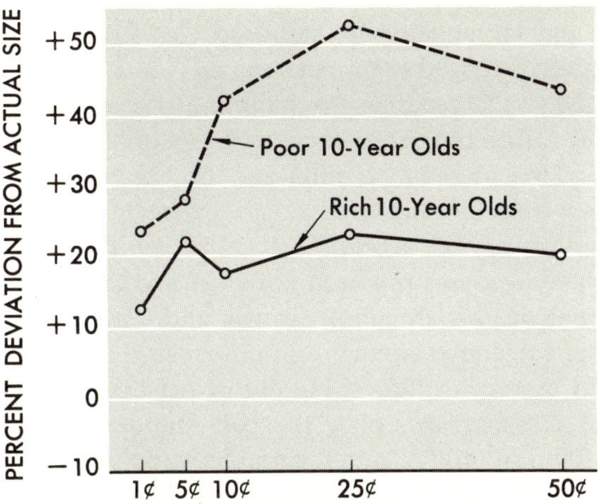

Figure 5.2 Judgment of size of coins by rich and poor children. From Bruner and Goodman (1947).

fifty cent coin and its cardboard disc. Bruner and Goodman suggested in their discussion of this finding that the inversion may have been due to the children's relative unfamiliarity with the fifty cent coin. Not being as familiar with the fifty cent coin, the children were less likely to be influenced by its value in judging its size. If they are correct in this interpretation, then it may also have been true that the poorer children were less familiar with all coins than were the richer children. Yet, this would suggest the expectation that the poorer children would have shown *less* rather than *more* overestimation of the size of coins. Obviously, experience with the valued objects must be controlled in an experiment of this type.

The Bruner and Goodman experiment had its origin in a transactionalist view of perception, as has been indicated. Analysis of this experimental problem in the context of a different theoretical system, that of multiplicative-drive theory (Brown, 1961), illustrates the point that substantial differences are found in the manner in which empirical testing might ultimately be undertaken. As Brown has observed, the drive theorist assumes that any influence of value or need factors in a perceptual situation would result from the interaction between energizing motivational variables (drive) and learned response tendencies (saying "smaller" or saying "larger"). Thus, rather than using the method of average error to study this problem, an effort should be made to employ a technique, such as the method of constant stimuli, from which both difference thresholds and points of subjective equality emerge as dependent variables. Using nonvalued cardboard discs, for example, subjects could compare a standard-sized disc repeatedly with discs of different sizes (both smaller and larger than the standard), each time reporting which disc looks larger. From the proportions of times each variable sized disc is reported to be larger than the standard, both a difference threshold and a point of subjective equality could be obtained. The threshold would represent the amount of difference in size between the standard and comparison stimuli which was reliably detected, while the point of subjective equality would represent the size of the variable stimulus which, on the average, was reported equal in size to the standard.

It is clear that, in this situation, Bruner and Goodman would predict that the point of subjective equality obtained using a coin as the standard stimulus should be greater than the point of subjective equality obtained using a disc as the standard, since the coin standard should look larger than the disc standard and a larger variable stimulus should be required to be seen as equal to it. No prediction could be made by Bruner and Goodman regarding the difference thresholds when coins or discs were used as standard stimuli, however.

Drive theory, on the other hand, would interpret a difference between the points of subjective equality obtained with valued and nonvalued standard stimuli as an associative rather than a motivational effect, since the only influence of its energizing factor would be expected to be a change in the steepness of the psychophysical function and, consequently, a reduction in the difference threshold in the higher motivational condition. The associative mechanism that might be assumed would be that of a general learned tendency to report more valued objects as "larger," acquired somehow in an experiential world in which "more" and "better" are frequently closely paired temporally.

Experimental Plan

Pursuing the foregoing approach to the problem of the perceived size of valued objects reveals additional significant aspects of the Bruner and Goodman study that might otherwise be overlooked. In the same way that problem analysis previously suggested that coins of several values be used in the Bruner and Goodman experiment, more than a single kind of coin could be used in the revised experiment. For example, suppose that both twenty-five and fifty cent pieces were to be used as standard stimuli, along with nonvalued standards of their respective exact sizes. A 2×2 experimental plan, shown in Figure 5.3, can then be used to represent the manner in which the four different experimental conditions are arranged.

Under each condition, the subject would look at a standard stimulus and a variable stimulus, side by side, and indicate which was larger.

	SIZE OF STANDARD STIMULUS	
	Small (25¢ size)	Large (50¢ size)
Valued Standard	25¢ piece	50¢ piece
Nonvalued Standard	25¢ size disc	50¢ size disc

Figure 5.3 A 2 × 2 experimental arrangement for studying the effect or value on perceived size, using the method of constant stimuli.

Each variable stimulus would be paired with the standard stimulus several times, and their locations on the left and right would be counterbalanced. Each of the four experimental conditions shown in the table would generate a point of subjective equality and a difference threshold for each subject in the experiment. Bruner and Goodman's theory would predict that the points of subjective equality for both valued standards (top row) should be greater than those for the two nonvalued standards (bottom row). In addition, this effect should be greater for the more valued standard (right-hand column) than for the less valued standard (left-hand column). No prediction is generated from the Bruner and Goodman theory regarding the difference thresholds. Brown's version of drive theory would imply that smaller thresholds should be found for the two valued standards (top row) than for the two nonvalued standards (bottom row). In addition, this effect should be greater in the right-hand column than in the left-hand column. Brown would make no prediction regarding the point of subjective equality, other than that no *motivational* factor could influence it.

Since the use of coins requires that the two conditions in the right-hand column of the figure involve a *larger*-sized standard stimulus than the two conditions in the left-hand column, the well-known Weber-Fechner law (Guilford, 1954) suggests that both of the right-hand conditions should show *larger* difference thresholds than the two left-hand conditions, regardless of the value of the stimuli. Since the Weber-Fechner law predicts larger difference thresholds for the larger standards, while drive theory predicts *more* of a difference between the valued and nonvalued thresholds in the two larger stimuli than in the two smaller stimuli (because the valued standard in the latter case is more valued than the valued standard in the former case), a rather complex interaction between the value and size variables might be expected to emerge from a study of this type, assuming that both processes are at work in this situation.

A recent study by Landsman (1964), using this experimental approach, found that neither difference thresholds nor points of subjective equality were influenced by the value of the coins that served as standard stimuli. Coin–metal discs served as standard stimuli in the nonvalued conditions of this experiment. That Landsman's procedure was sufficiently sensitive to detect subtle perceptual influences was attested to both by his finding that differences thresholds *were* significantly larger when larger standard stimuli were used (i.e., the Weber-Fechner effect was found) and by the fact that the point of subjective equality was significantly larger when nonvalued coin–metal discs were compared with nonvalued aluminum discs as standard stimuli (probably due to differences in reflectance of the two types of metal).

Regardless of what the eventual empirical assessment may be of the role of motivational variables, such as value and need, in perceptual processes, it is obvious that a great deal of methodological clarification can be achieved by careful analysis of what each relevant theory would predict in different experimental situations. Not only does it appear that Brown's suggested approach to the problem reveals the possibility that both motivational and associative mechanisms may be operative, but Landsman's extension of it to include the possible role of the well-established Weber-Fechner law makes it clear that the physical size of the coins used in these studies can be expected to have an effect over and above that which would be attributable to value or need.

EXPERIMENT ON HUMAN CONDITIONING

When human subjects are employed in studies of classical conditioning, observed changes in their tendencies to respond to the conditioned stimulus are sometimes attributed to nonassociative perceptual or response-set processes rather than to automatic "connections" between stimuli and responses. On one hand, the subject may form certain perceptual impressions regarding the patterning of events in the situation (e.g., "The airpuff always follows the red light but never follows the blue light"), while on the other hand, he may adopt a bias favoring either responding to the conditioned stimulus or nonresponding to it (e.g., "I will try to keep from blinking my eye in anticipation of the airpuff"). Several investigations have shown, in fact, that carefully worded instructions to the subject regarding either the stimulus situation or the way in which he should try to respond have observable effects on both acquisition and extinction of conditioning. While most of these studies have used responses mediated by the somatic nervous system (i.e., skeletal responses), which are under voluntary control, some have reported effects of instructions even when involuntary, autonomically mediated, responses are employed.

Factors Affecting the Instructions Effect

A recent study by Hill (1967) attempted to isolate one of the factors that might be responsible for the effect of instructions in human conditioning. This experimenter's analysis of the problem began with the assumption that instructions regarding how to respond in the situation could only be effective to the extent that the subjects were willing and able to follow them. Since following instructions may be assumed essentially to reflect compliance with the wishes of the experimenter, it follows that subjects who are highly compliant individuals would be more likely to show an effect of instructions than would subjects who are less compliant. This argument presupposes, of course, that the effect of instructions is primarily a matter of voluntary response control. On the basis of this reasoning, Hill administered a paper-and-pencil test of need for social approval, in order to identify two groups of subjects, one that was likely to be more conforming and compliant than average and one that was likely to be less conforming and compliant than average.

The main independent variable was a difference in instructions. This was implemented by urging half of the high-compliant and half of the low-compliant subjects to *try to become conditioned*, while the other half of each group was urged to *try not to become conditioned.* It was necessary, of course, to explain what conditioning is and to suggest that becoming conditioned involved accepting the conditioned stimulus as a sign that the unconditioned stimulus (an electric shock) was about to occur and preparing for it physiologically. Not becoming conditioned was said to require rejection of the conditioned stimulus as a basis for responding, since it was itself not noxious, and responding only to the shock.

It was expected that these pro- and anti-conditioning instructions would have opposite effects on the subjects' tendencies to make conditioned responses, resulting in a general difference in level of responding between the two instruction conditions. Of more significance, for our purpose of elucidating the use of problem analysis, was the expectation that this difference due to instructions would be larger in the two high-compliance subgroups. The reason that the latter expectation is especially important is that, if empirically verified, it would support the presupposition that the instructions effect involves the mediation of voluntary response processes. If the effect of the instructions variable were no greater in a highly compliant group of subjects than in a group which was low in compliance, the hypothesis that it is voluntarily mediated would be considerably weakened.

Suppression of conditioned responding in the anti-conditioning instructions condition might be the result of a discriminative process by which the subject distinguishes between the conditioned stimulus and the unconditioned stimulus and withholds the conditioned response until the latter occurs. Again presupposing that some kind of voluntary response control might be involved, Hill decided to introduce a third independent variable, the temporal interval between the onsets of the conditioned and unconditioned stimuli. Although only a simple conditioning paradigm was used in this experiment, it was reasoned that factors which have been shown to enhance discrimination (the CS-UCS interval) might also enhance the instructions effect. Thus, the four subgroups already considered (pro-conditioning instructions, high-compliance; pro-conditioning instructions, low-compliance; anti-conditioning instructions, high-compliance; and anti-conditioning instructions, low-compliance) were each divided in half to be conditioned with a long or a short CS-UCS temporal interval (500 msec. or 2000 msec.).

Experimental Plan

Figure 5.4 presents a diagram of Hill's experimental plan and indicates by means of hypothetical average conditioning scores, the expected results. On the left-hand side of the figure are the four groups conditioned with the 2000 msec. CS-UCS interval and, on the right-hand side are the groups conditioned with the 500 msec. interval. As is indicated by the hypothetical values in each table, the instructions effect was expected to be maximal in the high-compliance subjects who were conditioned with the 2000 msec. interval, since these compliance and CS-UCS interval conditions were both expected to facilitate the voluntary process by which the instructions variable was assumed to operate. At the 2000 msec. interval, the instructions effect was expected to be reduced in the low-compliance subjects. Similarly, even the high-compliance subjects were expected to show a reduced instructions effect when the CS-UCS interval was 500 msec. The difference between the pro- and anti-conditioning instructions was expected to be minimal (or even absent entirely) when both the CS-UCS interval and compliance conditions were unfavorable. No assumptions were made, in constructing this figure, regarding the *overall* effects of either the compliance or CS-UCS interval factors.

The increased empirical risk attendant upon more than cursory problem analysis is exemplified very well in this study, since its results were in considerable conflict with the major presupposition it attempted to evaluate. Both the instructions and compliance variables were found to effect the level of conditioning observed, the high-compliance groups responding at a higher level overall than the low-compliance groups and the pro-conditioning instructions groups responding at a higher level overall than the anti-conditioning instructions groups.

2,000 MSEC. CS-UCS INTERVAL

	Instructions	
	Pro	Anti
High Compliance	80	20
Low Compliance	65	35

500 MSEC. CS-UCS INTERVAL

	Instructions	
	Pro	Anti
High Compliance	65	35
Low Compliance	55	45

Figure 5.4 Diagram of experimental plan of study of pro- and anti-conditioning instructions, with hypothetical average conditioning scores illustrating expected results.

However, neither of the effects that were expected to reveal the voluntary basis of these differences was observed. The difference due to pro- and anti-conditioning instructions was no greater in the high-compliance subjects than it was in the low-compliance subjects, nor was the instructions difference greater when the long CS-UCS interval was used than when the short CS-UCS interval was used. In this case problem analysis made it possible to conclude that *both* the instructions and compliance factors have their influence on conditioned responding via some kind of involuntary mechanism, such as drive level, rather than being voluntarily mediated.

EXPERIMENT ON VERBAL LEARNING

In studies in which words are to be used as stimuli to which new associative responses are to be learned, it is very common for words of known association values to be chosen. Frequently this is done by selecting words from published lists which provide data regarding either the relative frequency of appearance of common words in the printed literature or the number and types of free verbal associations that are elicited by them.

On the assumption that even words with equivalent association values, according to such lists, may differ in their emotional characteristics and, thus, enter into verbal associations with different degrees of difficulty, Jacobs (1955) attempted to define word-classes on the basis of two types of quantitative reactions, the size of the galvanic skin response they evoked and the amount of time taken to react to them by saying another word. He then studied the formation of new verbal associations to words in the different classes. In a preliminary experimental session, words were assigned to one of four classes according to whether they produced large or small galvanic skin responses and whether they had short or long reaction times, both measures being obtained in a single session. Figure 5.5 shows the manner in which the four word-classes were constituted.

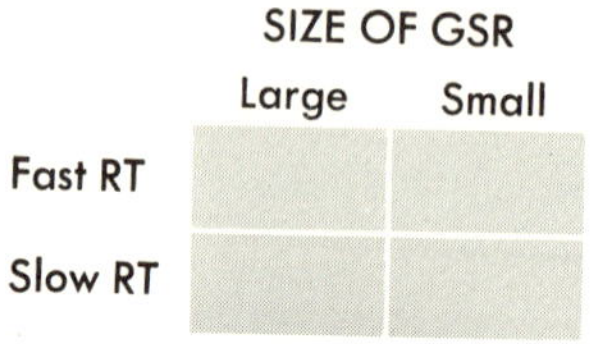

Figure 5.5 Four-fold classification of types of words based on GSR-RT combinations.

Complexity of Reaction Time Measures

The reaction time measure was assumed to reflect only imperfectly conventional association value, namely, long reaction times indicating weaker association values than shorter reaction times. Jacobs reasoned that some words with relatively high association values might require longer reaction times nonetheless, if they were emotionally charged and produced avoidance reactions. Even two words with equally long reaction times might differ in emotionality and, thus, in actual association value. It was assumed that the galvanic skin response elicited by the word would reveal its emotion-producing properties.

Even though an independent variable whose four categories are defined by inference from subjects' responses was used in this experiment, this does not relegate it to quasi-experimental status. Systematic deliberate manipulation of the types of stimuli employed was achieved by the experimenter, even though the differences among the stimuli were defined according to subjects' responses to them. This is not quite the same as dividing *subjects* into groups on the basis of their responses, for example, on a paper-and-pencil test.

Positive and Negative Feedback

Pursuing the notion that differences in the galvanic skin responses produced by the words would depend upon their capacity to produce emotional or avoidance reactions, Jacobs added another independent variable that was intended to establish different levels of emotional drive in different groups of subjects. Such a factor would, then, serve to amplify the tendency for emotional reactions to the words to differentiate them during the verbal learning part of the experiment. Half of the subjects were run in the learning part of the experiment under conditions intended to increase their emotional tendencies, since they were told repeatedly that they were doing poorly and were failing in the situation. The other half of the subjects performed without these deprecatory interjections.

This feature of Jacobs' problem analysis is quite similar to that employed in Hill's experiment on human conditioning, since in both cases an instructional variable was added on the assumption that its effect would tend to verify the assumed mode of operation of another independent variable. Jacobs expected that the extent to which emotional reactions to the words would interfere with the acquisition of new associations would be greater in the subjects who had been told that they were doing poorly than in the subjects who had not experienced this criticism, since he also assumed that the differences between the galvanic skin responses elicited by the words reflected emotionality and that emotionality would be enhanced by the failure instructions.

A greater number of correct associations were made to words with short reaction times than to words with long reaction times, as was anticipated. However, this difference was found to be independent of whether the words had elicited large or small galvanic skin responses. In fact, no systematic tendency for a difference in difficulty of association was found between the large and small galvanic skin response words.

Even more damaging was the finding that the instructions variable had no substantial influence on learning in the case of any of the four types of words. In fact, there was a slight tendency for the subjects who had been told they were failing to perform better than those who were not. Thus, there was no support for the assumption that the galvanic skin response evoked by a word either indicates the emotional reaction it produces or implies that the word will be more or less effective as a stimulus in a learning task.

However, Jacobs also had the words rated as to emotionality by a group of clinical psychology trainees and this method of defining emotionality effectively segregated those words which entered associations more and less easily. His analysis of the verbal learning situation, thus, was not entirely without cogency. As a further extension of this avenue of his problem analysis, Jacobs administered a personality inventory to the subjects some time after the conclusion of the verbal learning experiment and found that a selected group of subjects who had "blocked" on an emotional word during the experiment had considerably different and more maladjusted personality profiles than the remaining subjects who had not "blocked" on an emotional word. Blocking was defined as making no more than one correct response during the entire twenty trials of training.

The results of this study indicate that Jacobs' more elegant manifestations of problem analysis, which resulted in the four-fold classification of types of words and in the addition of the instructions variable, proved to be essentially fruitless. However, his extension of the analysis to include an independent measure of emotionality of the words (the ratings of clinicians) and his creativity in administering the personality inventory, illustrate the way in which the most useful features of experiments emerge from pre-experimental consideration of a variety of the possible implications of a given theoretical position. It goes without saying that, in the case of Jacobs' experiment, post-experimental reanalysis for the purpose of further study is absolutely essential. Certainly, his original idea of using reaction time and galvanic skin response measures to classify words appeared to have a priori merit. Careful consideration of the failure of this analysis to reveal expected effects might result in modifications in the experimental procedure that could, conceivably, even salvage the original format.

As Jacobs, himself, noted, when a protracted time between presentation of a word and the subject's reaction occurs, for whatever reason, the subject may become embarrassed and, as a result of this, make a large galvanic skin response. Thus, a low association value word which is not particularly emotion-producing, may be misclassified as highly emotional on the basis of a large galvanic skin response. Confining attention to only very short-latency galvanic skin responses might eliminate this problem and produce a more effective classification of words, since such responses could not be the result of the subject's embarrassment over the slowness of his verbal response.

Another way in which this problem could be handled would be to obtain measures of the galvanic skin response produced by the words when the subject had no task to perform other than merely looking or listening. The reaction time measures could then be obtained from a different sample of subjects whose galvanic skin responses were not being measured or from the same subjects on another occasion.

EXPERIMENT ON AVOIDANCE LEARNING

No characteristic of the behavior of complex organisms has received greater experimental and theoretical attention than their demonstrable ability to learn to prevent the delivery of a noxious stimulus by performing some response. Literally thousands of experiments have been done on this topic. Stripped of its complexities, the reason for this persistent interest in avoidance learning may be found in the fact that the subject learns to perform a response which prevents the occurrence of the very event assumed to be motivationally responsible for the acquisition of the response. Once the avoidance response is acquired and the subject no longer receives the original noxious stimulus, why does the tendency to make the response not simply go away? The simplest answer to this question is that the response remains adaptive in that its consequences are more satisfying than those of nonresponding. This answer is only fleetingly attractive. Its inadequacy is revealed by noting that prevention of the shock is a satisfying consequence only to the extent that shock has been experienced repeatedly and is "expected" imminently. Even if we accept the law of effect, we must still be concerned with the manner in which the subject learns to anticipate the shock in the first place, before he can learn to prevent its occurrence.

The avoidance paradox provided the major impetus for what has come to be known as "two-factor" learning theory (Rescorla and Solomon, 1967). As applied to the original avoidance paradigm, in which a neutral conditioned stimulus precedes an electric shock which can be prevented by the performance of the conditioned response, the two-factor theory states that the subject learns first to fear the conditioned stimulus and, only afterward, learns to reduce his fear by making the avoidance response.

The two-factor theory is, thus, also a two-stage theory. In the first stage, fear is conditioned to the conditioned stimulus by virtue simply of the close temporal contiguity of the latter with the unconditioned reflex to the shock. Here the "factor" responsible for learning is that of temporal closeness of the two experimenter-controlled events. In the second stage, anticipatory performance of the escape response to the unconditioned stimulus has the effect of terminating the conditioned stimulus and preventing the unconditioned stimulus. Here the "factor" responsible for learning is the reduction in fear achieved when the conditioned stimulus is foreshortened by the avoidance response. It is as if the subject were merely *escaping* from the fear stimulus much as he would from the shock itself.

The question has been raised whether it is necessary to invoke an essentially unobservable emotional reaction in order to explain avoidance learning. There is little doubt that it plays a role of primary importance in two-factor theory, since it provides both the motivational framework within which the avoidance response is initiated and the change in motivational state which allegedly maintains the avoidance response after it occurs. As is frequently the case in controversies of this type, appeal to the principle of parsimony is made, the argument being that it is theoretically inelegant and pragmatically inefficient to invent hypothetical processes to explain phenomena which can be comprehended at a less conjectural level of discourse.

As in almost all of these discussions, unfortunately, the question becomes one of agreeing upon the essential elements of comprehension. For example, some writers are less than satisfied with the orthodox two-factor position because of the burden it places on hypothetical emotional reactions. They prefer to speak of "aversion" attaching to the conditioned stimulus, operationally defining aversion in relation to the fact that removal of the conditioned stimulus (and the aversion) maintains the avoidance response. This approach is quite similar to that which employs the "empirical law of effect" in connection with appetitive instrumental learning. In the latter area, a "reinforcer" is defined as a response-contingent event whose effect is to increase or maintain responding. The presentation of such an event is, then, referred to as "reinforcement."

The difficulty with either of these approaches lies in the fact that each is a circular attempt to substitute *description* for *explanation.* The statement that aversion transfers to the conditioned stimulus and is removed by the avoidance response, tied as it is to a definition of aversion in terms of the response-maintenance effect of removal of the conditioned stimulus, is quite obviously totally circular and nonexplanatory. Explanation is *always* conjectural, *always* hypothetical. When total description is possible (if such a condition could even exist), explanation is precluded.

Free-operant Avoidance Learning

A more productive consequence of healthy skepticism regarding the two-factor theory is exemplified by a recent experiment (Herrnstein and Hineline, 1966) employing a technique aimed at demonstrating avoidance learning in the absence of a regular conditioned stimulus. These experimenters developed a procedure that extended the Sidman (1953) avoidance method to ensure that not even a regular temporal interval between responses and shocks existed.

In the Sidman procedure, each response has the effect of postponing the shock for a fixed interval, referred to as the response-shock interval. If the subject fails to respond during this interval, a shock is delivered and future shocks are programmed on a shortened shock-shock interval. For example, the shock-shock interval might be ten seconds, while the response-shock interval is sixty seconds. Even though there is no explicit conditioned stimulus which can acquire the property of eliciting a fear reaction, the two-factor theorists have observed that proprioceptive and other time-dependent stimuli can serve as elicitors of fear. Thus, they would argue, as the time since the last response increases, stimuli come increasingly into play that have been associated with previous shocks. These stimuli evoke greater and greater fear as more time passes. The avoidance response, by restoring stimuli (i.e., response-produced) that have never been closely paired with shock and by removing those that have, is reinforced by fear reduction. The fact that experiments using the Sidman technique have shown that the probability of response increases as a function of time since the previous response is taken as support for the notion that conditioned fear underlies even this signal-less avoidance behavior.

In Herrnstein and Hineline's procedure, the shock-shock interval is randomly varied around a mean which is shorter than the mean of the randomly varying response-shock interval. A response simply changes the mean rate of shock delivery to a longer value, but does not preclude the delivery of shock. As soon as the next shock occurs on the new schedule, the program is returned to the higher probability shock-shock schedule and another response is required to shift it back temporarily to the lower probability schedule. The fact that successful "avoidance" is generated by this procedure, in the sense that the subjects tend to respond so as to reduce shock frequency, attests both to the remarkable versatility of complex organisms and to the effectiveness of this example of problem analysis in zeroing in experimentally on a theoretical problem.

As Herrnstein and Hineline point out, the two-factor theorists could still claim that post-response stimuli have had a lower probabilistic association with shock than pre-response stimuli have had and, accordingly, the latter could serve as conditioned stimuli for the fear reaction and the former could mediate fear-reduction. But, since the relationship between these "fear" stimuli and shock is entirely a *probabilistic* one, it may be that parsimony is better served by assuming that the subject simply is sensitive to differences in shock frequency, as such, rather than assuming that conditioned fear still operates in the situation.

It is not so much that Herrnstein and Hineline's analysis of the avoidance conditioning problem discredits the two-factor theory as it is that their analysis drives the advocates of this theory into a highly vulnerable position, since little if anything is actually known regarding the conditioning of fear under the random shock schedule they have invented. A possible resolution of this issue may be that a conditioned stimulus is a sufficient but not a necessary condition for the establishment of avoidance behavior.

CHOOSING THE LEVELS OF THE INDEPENDENT VARIABLE

In Chapter 2 it was pointed out that the transition from simple hypothesis-testing experiments to experiments seeking to determine the quantitative nature of the relationship between the independent and dependent variables requires that more than two levels of the independent variable be used. This is due to the fact that two data-points can only reveal that one is higher or lower than the other, and nothing more. In order to determine more about the *shape* of the functional relationship between the independent and dependent variables, three or more points must be identified.

In certain types of experiments it is possible to predict in advance of data-collection only the most gross aspects of the shape of the functional relationship. In such cases it is necessary to sample a broad range of levels of the independent variable and to employ equal quantitative steps from level to level, assuming that a quantitative variable is being manipulated. For example, in studies of the shape of the generalization gradient (e.g., Grice & Saltz, 1950; Guttman & Kalish, 1956), five or more levels of the independent variable are often used, in an arithmetic series of steps.

Sometimes, however, prior knowledge regarding the effect of the independent variable in related studies provides a basis for using unequal steps along its scale. A geometric progression of numbers of training trials, for example, 3, 9, and 27 (Morrow, 1966), is effective in learning studies because so many previous studies in this area have obtained a negatively accelerated growth curve. To ensure that each successive increment in the independent variable will be reflected by an approximately equivalent behavioral increment, the increments must be increasingly larger.

A similar state of affairs exists in studies involving perceptual responses to simple stimuli, in which some intensive feature of the stimulus is varied systematically. Thus, to ensure that a wide range of perceptual response variation will be observed, a geometric series of stimulus variations may be needed, since the subject's perceptual response varies in proportion to the logarithm of the stimulus property varied. This relationship sometimes may be dealt with by transforming the subject's reaction, as was noted in Chapter 3.

PROBLEM ANALYSIS AND THE CRUCIAL EXPERIMENT

An experiment is said to be *crucial* if it succeeds in pitting two opposed theoretical explanations against each other in such a way that the results will definitively establish which is correct and which incorrect. With this definition in mind, experimenters are well aware that there is no such thing as a crucial experiment. In the light of the foregoing discussion of problem analysis in the planning of experiments, a more accurate statement might be that an experiment succeeds in being crucial to the extent to which the development of theory and the application of problem analysis have permitted. From this point of view, the "crucialness" of an experiment is one of its essential quantitative, descriptive properties. The more crucial the experiment is, the more likely that its results *will* settle a question unambiguously and enable the scientist to move on to another.

Even if problem analysis is pursued with great perspicacity and imagination, however, an experiment may seem to be less than crucial if the level of knowledge in the area is not high and competition among competing theories, accordingly, is absent. It is only after considerable accumulation of facts has required their articulation and theoretical systematization, that explanatory issues become sufficiently intense as to motivate "crucial" experiments of the historical type. More often, what is crucial is the job of making certain that simply stated questions can be simply and unambiguously answered experimentally. Once it is recognized that thorough problem analysis is required in all experiments, regardless of their theoretical significance, more experiments will terminate with definitive conclusions. Under such circumstances, it might even be discovered that "crucial" experiments are no longer necessary.

SUMMARY

A major aspect of *problem analysis* involves *choosing* the independent and dependent variables so that the broad goals of the experimental question may be achieved. The basic principle upon which this procedure is based is that the implications of the experimenter's tentative explanatory generalization can be realized fully only if he begins by assuming that it really is correct. Only in this way can he generate the kinds of empirical tests required for sensitive evaluation. In some experiments, this type of problem analysis results in the addition of independent variables that serve to extend the range of evaluation of the main independent variables (*dimensional analysis*) and that place his theoretical ideas in greater empirical jeopardy. This procedure may also sometimes have the beneficial effect of identifying sources of contamination of interpretation of the results.

Approaching the same general experimental question from two or more different theoretical viewpoints frequently results in strikingly different empirical analyses, even to the point of generating extremely different methods for manipulating the independent variables and for defining the dependent variable operationally. This may sometimes be the reason that different experimenters arrive at quite different answers to the same general experimental question. Yet, these differences in approach do not merely serve to produce confusion, since each different avenue of empirical attack on the same general theoretical proposition may contribute to the overall evaluation of it.

In research with human subjects psychometrically defined *quasi-independent variables* and *task* or *instructions variables* are often used to *amplify* (or otherwise modulate) the effects of the principal variables of interest, so that their underlying mechanisms of function may be better understood. Again, this device may sometimes reveal that the experimenter's own theoretical preconceptions were less than completely sound.

Problem analysis also contributes to decision making regarding the *levels of the independent variable* to be used. If there is good a priori reason to believe, for example, that the relationship between the independent and dependent variables is *nonlinear,* it may be desirable to manipulate the independent variable in geometric or logarithmic steps as well as to use at least three levels. Only by this approach can the effect of the independent variable be fully described and accurately interpreted.

6

Significance of Experimental Results

PRINCIPLE OF RANDOMIZATION

Experiment on Dermo-optical Perception

In the last few years there have been several reports in the Russian literature of studies of individuals who allegedly have the ability to perceive visual information through the skin, simply by placing their fingertips or palms over the stimulus material. Such "eyeless vision" is, to say the least, the topic of a spirited debate among Russian and some American psychologists. The debate focuses on the basic question of whether dermo-optical perception is, in fact, a scientifically demonstrable phenomenon, as well as touching on the matter of its extent and whether or not it requires training. The most famous case of dermo-optical perception, Rosa Kuleshova, allegedly read ordinary newspaper type at the rate of 170 characters per minute, one-fifth her normal reading rate with her eyes, and errorlessly distinguished all colors, in both instances using her right index and middle fingertips only (Dobronravov & Fishelev, 1964).

If we were called upon to conduct an experiment to determine whether an individual who is supposedly able to identify colors using only his fingertips actually could do so, how might we go about doing the experiment? Suppose, for simplicity, that we decided to use only two very distinct and spectrally separated colors, red and blue, and that appropriately dyed test patches were available for this purpose. The subject is securely blindfolded and permitted to place his index and middle fingertips immediately above the surface of a test patch, but not in direct tactual contact with it. He is to report whether the color of the patch is red or blue.

If the subject does not have the ability but merely guesses, he will be correct half of the time, on the average. This is so because there are only *two* possible outcomes and *one* of them is correct. It is clear, therefore, that more than a single test trial will be needed to satisfy us. The italicized phrase used two sentences previously, commonly known as the *null hypothesis,* permits us to elaborate in detail exactly what the likelihood is that behavior consonant with having the ability might occur if the subject actually does not have the ability but guesses. The null hypothesis enables us to do this because it provides a model of the relative frequencies of all of the possible outcomes of the experiment, including both those favorable to the hypothesis and those unfavorable to it.

Following the example of R. A. Fisher [1] in his famous experiment on the lady tasting tea (Fisher, 1953), we inform the subject that the experiment will involve eight trials and that they will consist of four patches of each color. He is told also that the order in which the stimuli are to be presented will be randomized (this will require an explanation of exactly what we mean by "randomized").

The null hypothesis allows us to specify exactly how often a correct division of eight stimuli into four red and four blue would be expected by chance, since it consists of a complete enumeration of all of the possible outcomes in this situation. Eight stimuli can be divided into two sets of four in seventy different ways ${}_8C_4 = 8!/(4!)(4!)$. But only one of these is in complete accord with the hypothesis that the subject can correctly distinguish the colors red and blue using only his fingers. Thus, if the subject correctly identified the colors of all eight test stimuli, we could state *with certainty* that this could only have happened by chance as often as once in 70 times. While this result may be described as statistically significant, in that *alpha* is smaller than the conventional 0.05 significance level, it is still far from certain that the subject has the claimed ability. To be sure, our stated significance level indicates the probability that he does not; and he may *not.*

This limitation notwithstanding, we would be inclined to conclude that the results of *this* experiment tend to confirm the hypothesis that he does have the ability, recognizing that we could never prove our experimental hypothesis beyond doubt even if we repeated the experiment with similar results many times. One can never be absolutely certain that chance was *not* responsible for the results, but the manner in which the null hypoth-

[1] While Fisher's plan is taken as a beginning point for this presentation, it will be made clear that a more sensitive alternative may be preferable.

esis is stated enables us to be quite sure of the *chances* that our conclusion is in error.

The order in which the red and blue test stimuli are presented to the subject is randomized, as was noted above. Also probably randomized, but not previously noted, would be the different pieces of material which receive the different colors (i.e., which patches are colored red and which blue). Randomization ensures that those factors that may influence the observed behavior, other than the ones under study, will have an equal opportunity to have an effect under all of the experimental conditions. It is the principal experimental device for dealing with error that is not reduced or controlled by other methods. In the dermo-optical perception experiment, the random ordering of red and blue trials guarantees that each trial may be of either type by chance. Any factors which change progressively from trial to trial will have an opportunity to affect the subject's judgments of red and blue only in a random fashion. Randomizing the assignment of colors to different pieces of material permits only chance to dictate the extent to which the irrelevant cues of size, shape, etc., may influence the behavior. Only because randomization has been used in this way is the null hypothesis appropriate. For this reason randomization is referred to as the physical (i.e., experimental) basis of the null hypothesis.

Distributing Error Randomly

It is important to understand that the purpose of randomization is *not* to equate the experimental conditions with respect to the values of the randomized variable, but *to distribute them randomly.* Instead of randomizing the order of presentation of the two types of stimuli, for example, it would have been possible to employ some kind of balanced sequence of trials which ensured that both colors would occur with equal frequency in the early and late trials. Similarly, it would have been possible to identify the various ways in which the different pieces of material to be colored tended to differ beforehand and, based on this information, either to arrange for each color to be associated equally often with each different size, shape, etc., of material or to use only a single, unvarying type of material (i.e., to hold these factors constant). Even in the latter instance, however, very small variations would probably still be present.

Even after differences in material have been reduced to some feasible minimum, those differences that remained would still have to be randomized, just as differences in order that remained after early–late balancing would have to be. No matter how much effort is expended in attempting to reduce variation by balancing, matching, holding constant, or by the elaborate error-reduction methods presented in Chapters 8, 9, and 10, randomization must finally be used to provide a basis for establishing a chance null hypothesis.[2]

A more sensitive experiment, in fact, can be conducted on the dermo-optical perception problem if we abandon the procedure of equating the number of red and blue stimuli to be used. Instead of telling the subject that there will be an equal number of red and blue stimuli whose *order* of presentation is randomized, we can tell him that the *selection* of a red or blue stimulus will be done randomly on each trial. No additional explanation will be necessary for the subject (compared to the earlier form of this experiment), since it is probably no easier for him to understand how the order of presentation of four red and four blue stimuli would be randomized as how a random selection of red or blue would be made on each trial. (The latter, in fact, is likely to be somewhat easier to communicate, since it can be explained that a coin will be tossed prior to each trial for this purpose. This is also somewhat easier for the experimenter.) Using four stimuli of each color is no less a form of balancing than is using different kinds of material equally often for each color or having each color occur equally often early as late in the sequence of trials.

Experimental Task Determines the Null Hypothesis

Because the problem is no longer one of dividing a set of stimuli into two equal subgroups, a somewhat different null hypothesis is generated. Each presentation of a stimulus to be identified as red or blue is a fresh 50–50 proposition, both for the subject and for the model of chance outcomes. The probability of identifying all eight stimuli correctly now must be determined in the light of the fact that any particular sequence of eight stimuli used is but one of 256 that are possible, instead of the 70 in the previous version of the experiment. Outcomes ranging all the way from all eight red to all eight blue can occur, rather than their being restricted to sets containing four of each color (i.e., the total number of outcomes is now 2^8). A subject without "eyeless vision" would *guess* all eight stimuli correctly only once in 256 times, on the average. Simply by changing the procedure as indicated (from four of each color to random determination of color on each trial), the same outcome (all guesses correct) is now far more significant.

This increase in sensitivity is only reasonable, since the subject's *guessing* task in the revised experiment is far more difficult than it was in the first experiment. However, if he really has the claimed perceptual ability, both tasks should be equally easy! Since the amount of time and effort

[2] The economical principle mentioned in Chapter 2 in connection with the size of sample used in an experiment applies here as well. Our object is not to eliminate all error, but only to reduce it within manageable limits and then to randomize.

required of both the experimenter and the subject is the same in both experiments, the latter would seem to be preferable. There does not seem to be any persuasive reason for using half red and half blue stimuli, since a subject who could identify, say, all red ones could identify all nonred ones as easily (given that he has been informed that only red and blue will be used). Furthermore, the null hypothesis in the second situation also seems logically to be more appropriately contrasted to the claimed ability.

Now, it may be that the person has the alleged ability but, because it is a less acute perceptual skill than using the eyes, perfect performance seldom would be achieved under the trying conditions of our experiment (just as a person with perfect vision might occasionally misidentify visual stimuli under impoverished stimulational circumstances). Suppose that only three red and three blue stimuli were correctly classified in the first experiment (we will assume that one misclassification would result in two errors, because of the subject's knowledge that there are four of each color). How would we evaluate these results?

Under the null hypothesis of the first experiment, three correct red classifications and three correct blue classifications (along with one incorrect in each color) can occur in 16 different ways out of the 70 possible outcomes. In addition to these 16 outcomes favorable to the hypothesis, *perfect* performance must, of course, also be counted, since our problem now is one of determining how often by chance a performance of six correct out of eight, *or better*, could occur. There are, thus, 17 chances in 70 that a person who does not have the claimed ability could achieve this level of success by guessing alone. We would not be willing to reject the null hypothesis under these circumstances, since its chances of being true are too great to risk.

By comparison, the revised version of the experiment is more sensitive in cases of less than perfect performance, just as it is when perfect performance occurs. Because each single trial involves a 50–50 chance of guessing correctly, regardless of the outcome of any previous trials, the revised experiment permits a single error to occur (without the necessity of an opposed error, as is the case in the original experiment). With a 50–50 chance of guessing correctly on each trial, two successive correct guesses would occur once every four times, three successive correct guesses once every eight times, etc. Seven correct and one incorrect responses can occur in eight different ways out of the 256 possible outcomes of the revised experiment. Adding to this the one way in which perfect correct performance can occur, the chance likelihood of being correct in seven out of eight trials, or better, is only nine in 256. The null hypothesis would still be rejected at a conventional significance level. Even a performance of six correct out of eight, or better, in the revised experiment, could occur only 37 times in 256, a result whose chance probability of 0.14 is still lower than the probability of 0.24 associated with the same result in the original experiment.

Still another way of demonstrating the greater sensitivity of the second experiment as compared with the first is to ask what the minimum necessary number of trials would be, given perfect performance, to warrant rejection of the null hypothesis. As we have seen, a minimum of eight trials (four of each color) is needed in the first experiment, because a single error (with its reciprocally opposed error) has a chance probability of 0.24. In the revised experiment, on the other hand, as few as five trials would produce statistically significant results if perfect performance occurred. The chance probability of five correct responses in five trials is only one in 32, $(\frac{1}{2})^5$. Six correct responses in six trials (one in 64 by chance) and seven correct responses in seven trials (one in 128 by chance), of course, are also both significant results in the revised experiment. Even six correct, or better, in seven trials almost achieves a conventional significance level (eight in 128, or $p < 0.07$).

EXPERIMENTS INVOLVING CONTINUOUS MEASURES

Experiment on Self-Punishment

In planning his experiment, the researcher must anticipate and prepare for the arguments that will be necessary to establish the probability that chance alone could have been responsible for his results. He does this (1) by conceptualizing in advance how chance alone is likely to influence the dependent variable he has elected to measure and (2) by using randomization of all uncontrolled factors that may influence these measures. The manner in which randomization permits a statement to be made of the probability that chance alone can account for the results is somewhat different in experiments involving quantitatively measurable variables than in those in which simple frequencies of occurrence of qualitatively different responses are obtained. Nevertheless, randomization remains the operational basis upon which the logic of the null hypothesis is established.

Vicious Circle Behavior

Consider, for example, an experiment in which a rat is trained to run down a straight alley to escape from an electric shock and is then given test trials with the shock removed from all but a part of the alley (Melvin, 1964). The animal is now in a position to avoid shock altogether simply by not running when placed in the alley, since the beginning part of the alley into which the rat is placed is now not electrified. Nevertheless, the animal persists in running down the alley, bringing itself into c with the shock, and ending up in the unelectrified goalbox. The response measured in this experiment on self-punitive or *vicious*- behavior is the speed with which the rat traverses the entire alley.

The experimenter is interested in whether the presence of shock in a portion of the alley is in any way responsible for the persistence in running demonstrated by the rats. It is possible that animals given the preliminary escape training would continue to run down the alley rapidly even if no shock at all were present. One way to deal with this question is to use *two* groups of rats, one (the *experimental* group) to receive shock in a part of the alley following the escape training, and the other (the *control* group) to be run with no shock at all after the initial escape training. After all of the subjects have received a standard amount of escape training, half of them are tested with shock in part of the alley and the other half are tested without any shock. The division of subjects into experimental and control groups is done randomly. For the purpose of the present discussion, the dependent variable for analysis will be defined operationally as the average running speed of each subject during the test series of trials.

The experimental hypothesis predicts that there will be a difference, on the average, between the running speeds of the subjects in the two groups. If the speeds of the subjects in the experimental group manifest the effect of the shock in the alley and those of the subjects in the control group do not, we would expect a difference in favor of the former over the latter.

The hypothesis is focused on the difference between the *average* speeds of the two groups, rather than on the performance of individual subjects, because uncontrolled variation in the dependent measures makes any single observation an unreliable index of the effect of the different treatments. That is, a subject who typically runs very rapidly may do so even if he is in the control group, without any influence of the shock in the alley. Similarly, a subject who typically runs very slowly may continue to do so, relatively speaking, even if the shock is present. However, if such variations are due only to chance (where "chance" means both that subject differences are randomly distributed between the experimental conditions, as a result of random assignment, and that random factors underlie individual differences in running speed), their distribution is described by the *law of error*.

Assumptions Underlying the Law of Error

The law of error is a completely rational (mathematical, as distinguished from empirical) model describing the theoretical relative frequency of random errors of varying sizes and directions (i.e., upward or downward from the "true" measure). It is derived from two basic assumptions, that errors in either direction are equally likely (i.e., that their distribution is symmetrical), and that the probability of occurrence of an error (in either direction) is an inverse function of the size of the error. That is, very large errors occur only very seldom, while small errors occur more frequently.

The mathematical equation which represents the law of error is known as the equation of the *normal curve.* That this function is closely approximated by the natural distributions of a variety of biological and psychological measures is, of course, one of its main virtues, as well as testimony to the fact that many of the ways in which biological and psychological variation occurs are fundamentally random (e.g., the manner in which alleles separate and recombine in the reproductive process).

If a difference between the average running speeds in the two groups were found (and, if it were not due merely to chance), it could not be attributed to the simple experience of the preliminary training, since the subjects in both groups received equal amounts of escape training prior to the test trials. We would be inclined to conclude that the presence of the shock in the alley for the experimental group was the key factor in the difference between the responses of its members and those of the control subjects.[3]

Implementing the Law of Error

What specific steps must be taken by the experimenter in planning and conducting the experiment on the vicious-circle effect so that the law of error will be the appropriate model of the relative frequencies of all of the possible differences that might be found between the average speeds of the two groups? First, he must be certain that all of the factors that may influence the running speed of a subject in either group, other than those that are eliminated or held constant by control procedures (e.g., data will be collected in a regulated environment), and other than the

[3] The two groups also differ in this experiment in whether any stimulus at all is presented during the test series. Even if the experimental group displays a faster average running speed than the control group, this might be due to the presence of the shock as a *stimulus,* rather than because it is the same aversive stimulus from which the subjects previously learned to escape. To overcome this type of *confounding* of two possibly relevant factors, another control group could be run which receives a sound or a light at the same point in the alley at which the experimental group receives the shock. If the persistence in running in the experimental group is to be attributed to the presence of shock in the alley, as opposed to the presence of any stimulus, a difference in average speed must be found between the experimental group and the additional control group. The point could still be made that the experimental group's shock differs from the additional control group's stimulus in that the latter is a novel event. As is pointed out in Chapter 11, confounding, albeit unfortunate, is still sometimes unavoidable.

one which defines the independent variable (in this case, the presence or absence of the shock in the alley during testing), are randomized. Thus, for example, it may be arranged that subjects run at different times of the day, week, or month, have an equal chance of being assigned to either group. The experimenter will not run all of the subjects in the control group first, followed by all of the subjects in the experimental group.

It is likely that not only time of day or week but also changes over time in the laboratory instruments and even in the experimenter's procedural skill may influence the animal's running speed. Even if some procedure for balancing the distribution of occurrences of experimental and control conditions over time is used (such as running an equal number of experimental and control subjects each day), randomization of conditions within days must be employed. The extent to which this point is overlooked in many experiments is unfortunate, since balancing can only reduce, but not completely eliminate, error. Furthermore, that which remains can only be estimated if randomization is employed in conjunction with the balancing.

Random Assignment of Subjects

Most important of all, subjects must be assigned randomly to the different experimental conditions, since individual differences among subjects constitute the single largest source of error in psychological experimentation. As in the experiment on dermo-optical perception, random assignment of subjects to the two conditions is *not* intended to *equalize* them with respect to subject characteristics, but is done to enable us to evaluate our chance null hypothesis.

If there are differences among subjects that are known to be associated with substantial variations in running speed, it may be necessary to employ some kind of matching, or balancing, procedure to control the effects of these particular factors. Thus, it is quite common in animal and human studies to arrange that the proportion of males and females in the different experimental conditions be equated. Even when this type of sex-balancing is used, however, randomization is what determines *which* males and *which* females are assigned to the different groups, as well as in which order they will be run. Even though the most frequent outcome of random assignment of subjects to experimental conditions is a set of more or less similar groups, especially when group size is quite large, this is not the purpose of the procedure. If highly similar groups were sought, as sometimes will be the case, more effective methods than randomization would be used.

Our null hypothesis states that the average running speeds of the two groups should differ only by chance, just as would the means of random samples taken from the same population. When two samples of measurements are taken randomly from the same population (ideally, with *replacement,*[4] so that every measure in the population has a chance of being in either sample), two factors determine how different their means may be. The first of these is the size of the samples. The means of very small samples are more likely to differ from one another than are the means of very large samples. This should be clear simply by noting that the means of very large samples are much more likely to be close to the population mean than are the means of very small samples, i.e., if the sample were as large as the population, its mean would be identical to the population mean, while, if the sample had only one case in it, its "mean" would seldom be very close to the population mean. The greater the number of cases in the sample, the more "representative" of the population the sample is likely to be. Since large sample means are more likely to be close to the population mean than are small sample means, they are more likely to be close to one another as well.

In addition to the size of the samples, the other factor determining how different the means of two random samples drawn from the same population may be is how variable are the measures themselves in the population. The greater the variability of the parent population, the greater the variability among means of random samples taken from it. This should also be intuitively clear, since, as was implied above, if each sample consisted of only a single case, maximum variation among sample "means" would be possible (their variance would be the same as that of the population from which they came). If, however, each "sample" were as large as the entire population, all of their means would be the same (the variance of their means would be zero).

TESTING THE NULL HYPOTHESIS

These considerations provide us with a method for assessing the probability of obtaining by chance varying amounts of difference between sample means, since, under the null hypothesis (i.e., no difference in running speed is produced by the different procedures used in our experiment), the sample means are assumed to have come from the same population. Random assignment of subjects, randomization of times, etc., enable us both to assume a null hypothesis that says we are dealing with *random* samples from the same population and to obtain an estimate of the variance in the population from the variance of the measures of

[4] After each item is selected, it is placed back into the population and could, theoretically, be selected again.

subjects *within* the different groups. Since the subjects were assigned to the different groups at random, the groups' means should differ only in proportion to the extent that variability among individuals' running speeds exists naturally (i.e., in the population of subjects). Within each group, subjects differ in their responses even though they were treated similarly in the experiment, because of the same natural variability.

Using the within-group variance as a basis for estimating the population variance (called, in this context, the "error term"), we can comparatively assess the variability of the sample means (in our experiment, the *difference* between the two means). Mathematical statistics tells us that the variance of a distribution of sample means (all from the same population) equals the population variance divided by n, the size of the samples. With this knowledge, most of the standard statistical tests of significance of differences among sample means (e.g., the t test, the F test), can be comprehended. None of this could be achieved without proper implementation of the principle of randomization.

In the experiment on vicious-circle behavior which we have been considering, a t or F test of the difference between the obtained means would both be appropriate.[5] Since the F test lends itself for use in a much wider variety of situations, discussion of it will be more fully presented in Chapter 7. The t ratio, in the language of the preceding discussion, is formed by dividing the obtained difference between the two means by the "standard error" of the difference. The standard error is estimated by dividing an estimate of the population variance by the number of cases in the sample and taking the square root of this quantity. Except in the present situation, we have *two* independent estimates of the population variance, one from each sample. Since the null hypothesis would suggest that these are both estimates of the same population variance, a combined estimate of the population variance is used to obtain a standard error for each mean. These are then also combined to obtain the standard error of the difference between means. The basic formulas and their computational equivalents are presented in Table 6.1.

The probability of obtaining t ratios of varying sizes, *if the null hypothesis is true* (i.e., if the samples differ only by virtue of random sampling from the same population), is known exactly and is available in the Appendix. If our obtained value of t could have been obtained by chance only 0.05 or less of the time, we would reject the null hypothesis and consider the alternative hypothesis that the difference in the two groups' average running speeds may have resulted from the difference between the experimental and control procedures.

[5] In this situation t is equal to the square root of F and both test the same hypothesis.

TABLE 6.1

Basic and Computational Formulas for the *t* ratio.

1. $t = \dfrac{\bar{X}_e - \bar{X}_c}{s_{\bar{X}_e - \bar{X}_c}}$	The difference between means divided by the standard error of the difference between means.
2. where, $s_{\bar{X}_e - \bar{X}_c} = \sqrt{s^2_{\bar{X}_e} + s^2_{\bar{X}_c}}$	The standard error of the difference between two means computed from the separate standard errors of the two means.
3. and, $s^2_{\bar{X}_e} = s^2/n_e$; $s^2_{\bar{X}_c} = s^2/n_c$	Note that the same estimate of the population variance, s^2, is used in both of these, as explained in the text.
4. and, $s^2 = \dfrac{\Sigma x_e^2 + \Sigma x_c^2}{(n_e - 1) + (n_c - 1)}$	The population variance is estimated using the combined sums of squared deviations from the two samples.
5. and, $\Sigma x_e^2 = \Sigma X_e^2 - (\Sigma X_e)^2/n_e$; $\Sigma x_c^2 = \Sigma X_c^2 - (\Sigma X_c)^2/n_c$	Because: $x = X - \bar{X}$; $\Sigma x = \Sigma X - n\bar{X}$; $\Sigma x^2 = \Sigma X^2 - (\Sigma X)^2/n$

6. A convenient computational formula, thus, for determining t in the case of two independent (uncorrelated) means is:

$$t = \frac{\bar{X}_e - \bar{X}_c}{\sqrt{\left(\dfrac{\Sigma x_e^2 + \Sigma x_c^2}{n_e + n_c - 2}\right)\left(\dfrac{1}{n_e} + \dfrac{1}{n_c}\right)}}$$

The Error Estimate

Whether the null hypothesis is true or false, the within-group variations may still be used to estimate the population variance, since, even if the different experimental procedures result in different average running speeds, it is assumed that whatever increment or decrement is producd by the difference in procedure affects all the subjects within each group similarly. This is known as the assumption of *additivity,* i.e., that the effect of receiving a particular level of the independent variable can be conceived of in terms of a simple increment or decrement to all of the separate scores in the group, and its observable consequence is referred to as *homogeneity of variance,* i.e., the separate within-group variances all are estimates of the same population variance and should differ only by chance. It should be recognized that this assumption is fundamental to the procedure of combining data from different conditions to obtain a single (called "pooled") estimate of the population variance.

REPLICATION

Psychologists and statisticians use the term "replication" somewhat differently. When an experimenter has attempted to repeat an experiment done previously, his study may be referred to as a "replication," or, if the results obtained do not agree with those reported earlier, as an "attempted replication." On the other hand, statisticians define a replication as the smallest number of observations needed to provide a comparison among all of the experimental conditions. Even the first set of observations, one from each experimental condition, is already considered a "replication." Thus, in our experiment on the vicious-circle phenomenon, two subjects, one in each condition, would constitute one complete replication.

In this book, the term "replication" is used in a way which combines some of the features of both of these different usages. At least one repetition of the basic comparison implied by the experimental hypothesis is needed for "replication" to be accomplished. If the same subject receives all of the experimental conditions (a form of "blocking" described in Chapter 9), *two* subjects must be run for replication to be achieved. If these subjects differ only by random sampling from some population of subjects, these are called "random replications." On the other hand, if the hypothesis must be evaluated by comparing two subjects who are treated differently, *four* subjects will be needed for this basic comparison to be replicated.

In a somewhat more general way, the principle being developed is that the experimenter must replicate his observation of each phenomenon whose significance he wishes to evaluate. If his operational definition of the basic phenomenon involves a comparison of two subjects who are treated differently, two of such comparisons are needed to achieve replication, and these separate comparisons should, ideally, differ only randomly. If the operational definition of the phenomenon involves a difference between two differences, for example, when the experimental hypothesis predicts an interaction between two independent variables, the experimenter must provide for replication of this double-difference in order to be able to test its significance. If the experimenter wishes to test the hypothesis that an independent variable, administered completely within subjects, has a different effect on different subjects, the subjects will have to be measured at least twice at each level of the independent variable.

Replications that differ only by chance provide an "appropriate" error estimate. This is the most essential aspect of the application of the randomization principle in the planning of experiments. If one replication of the phenomenon under study differs from another (or others) only randomly, an estimate of chance variation *in the* (operationally defined) *phenomenon* can be obtained. If the experimenter accurately identifies his "phenomenon," and provides for its random replication, an appropriate error estimate will automatically be available for his statistical test. Most of the failures in experimental planning that result in the unavailability of an error term appropriate for testing an effect of interest can be traced to faulty conceptions of either how the basic phenomenon under study was operationally defined or what had to be done in order to replicate it properly.

In describing each of the principal experimental designs presented in this book, the manner in which the appropriate error terms for testing each effect are generated and how they can be identified will be reviewed. In recent years, the development of mathematical statistics into an independent discipline has had among its hallmarks the discovery that different mathematical models underlie the different experimental designs employed, and that proper error terms for testing various effects can be determined on the basis of mathematical principles.

A brief presentation of these models is given in Chapter 7, along with some "rules of thumb" for selecting appropriate error terms in the more complex designs, although it is hoped that even the nonmathematical reader will be able to arrive at an intuitive comprehension of the basic principle outlined above. In all but the most complicated experiments, the choice of the appropriate error term (given that one is available) can be made without resort to mathematical principles. Naturally, an experimenter who has thought about his experiment sufficiently to plan for random replication of the basic phenomenon under study would probably have little need for mathematical concepts in choosing his error terms.

SUMMARY

In an experimental situation in which the observed behavior can have only one of two possible values, and the chance probability of either value equals 0.5, it is necessary to replicate observations sufficiently to establish that the observed overall results reflect a real rather than a chance phenomenon. Assuming that chance alone is operating (the *null hypothesis*), a model is generated which describes the relative frequency of any and all possible overall outcomes. If the particular overall outcome could have occurred by chance only very rarely according to this model, the null hypothesis may be rejected. The actual chance probability of occurrence of the observed result (e.g., 10 straight identical observations, $p. < 0.001$) indicates the likelihood that rejecting the null hypothesis may have been an error.

In employing a *binomial* chance model of this type, all factors capable of influencing the experimental observations must be randomized if they are not controlled, since randomization is the physical basis of the null hypothesis. Aside from a possible real experimental effect, nothing but chance should be permitted to influence the results.

When the dependent variable measures are capable of a wide range of continuous quantitative variation, a more complex model of the null hypothesis must be employed, based upon the *normal distribution*. Nevertheless, randomization remains the essential basis of the null hypothesis test. In an experiment of this type involving an independent variable with two levels, all of the extraneous factors that may influence the dependent variable (other than those controlled experimentally) are randomized. Thus, for example, subjects are assigned at random to the two conditions. The goal of random assignment of subjects is not to equalize the groups but to provide a physical basis for the null hypothesis to be evaluated.

In the two-condition experiment with continuous measures, the null hypothesis states that the dependent measures in the two conditions *do not differ on the average*. If the null hypothesis were actually true, only two factors would determine how different the two averages might be: (1) the size of the samples and (2) how much variation actually exists in the dependent measures generally. The means of very large samples of observations are likely to be more similar than those of small samples. Likewise, the more variable the dependent measures are generally, the more different should be the means of two only randomly different conditions.

The actual test of the null hypothesis involves a comparison of the observed difference between the means of the two conditions, on one hand, and an estimate of the standard deviation (a standard measure of variability) of the distribution of randomly sampled means, on the other. On the basis of known probabilities of obtaining mean differences of varying standard amounts (e.g., the *t* distribution), the null hypothesis may or may not be rejected.

The essential feature of both of the foregoing types of null-hypothesis testing is that *randomization* and *replication* be employed appropriately. Randomization refers to the manner in which extraneous factors (error) are distributed, and replication refers to the repetition of observations over which this distribution is achieved. To have an *appropriate estimate of error* for evaluating the significance of an experimental result, the experimenter must have arranged for random replication of the *phenomenon* under study. This, of course, requires that he identify the operational definition of the phenomenon so that its replication may be accomplished.

7

Statistical Inference

LOGIC OF THE *t* TEST

In Chapter 6 it was noted that the t ratio for testing the significance of the difference between two sample means is formed by dividing the obtained mean difference by an estimate of the standard deviation of a hypothetical distribution of such mean differences, the "standard error" of the *difference* between means. The distribution of values of t which would occur by chance for samples of different sizes (from the same population) is known, so that the probability of obtaining t's of different values by chance may be determined. The method developed as an outgrowth of a more general procedure that employed the normal distribution and used z ratios rather than t ratios.

Standard Scores

A z ratio, or z score, is obtained by dividing the difference between a particular value of X and the mean of X by the standard deviation of X. In its most simple uses it is merely a device for converting X into a "standard score," one which expresses X in terms of relative amount of deviation from the mean. Obviously, comparisons among X's may sometimes be more meaningful if they are expressed as standard scores. This is the method used by Galton in his studies of the relationship between various characteristics of parents and their offspring, studies which resulted in the development of the Pearson correlation coefficient (which is defined as the average of the cross-products of the z scores).

When X is itself normally distributed $(X-X)/\sigma$ is also normally distributed, since the z score is simply a linear transformation of X. This normal distribution has a mean of zero and a standard deviation of 1.0, because of the manner in which the scores are transformed. If a single score were selected randomly from a normal distribution with known mean and standard deviation, it could be converted to z and the probability of its occurrence (more precisely, the probability of obtaining a score equally or more deviant from the mean) could be determined from the tabled probability values of the normal distribution. For example, a z of ± 2.0 or greater can be expected to occur by chance only about 5% of the time, since approximately 95% of the area under the normal curve is contained within ± 2.0 standard deviations. A table of areas under the normal curve for various values of z is given in the Appendix.

The logic of this procedure applies to means as well as to individual scores, except that the quantity to be divided into the deviation must now be the standard deviation of *means.* In other words, just as $X - \bar{X}$ must be divided by the standard deviation of X to obtain a standard score, $\bar{X} - \mu$ (the population mean, which is the mean of sample means) must be divided by the standard deviation of $\bar{X}$ to obtain a standard score. Since the standard deviation of $\bar{X}$, denoted by the symbol σ_X equals $\sigma / \sqrt{n}$ (the population standard deviation divided by the square root of the size of the sample which produced $\bar{X}$), the chance probability of obtaining a particular sample mean (or one even more deviant from the population mean) may be determined whenever the population mean, μ, and the population standard deviation, σ, are known, assuming that the distribution of $\bar{X}$ is normal.

Central Limit Theorem

Mathematical statisticians have shown that the distribution of sample means approaches normality as the size of the samples increases, regardless of the shape of the original distribution of X (the *central limit theorem*). The only serious limitations of this procedure are the requirement that the population mean and standard devivation be known and the need for using moderately large samples.

However, the experimenter actually *assumes* the value of the population mean when he states his null hypothesis. For example, a particular null hypothesis may be that when a subject is observed under the two ostensibly but not really different conditions no consistent difference in performance should occur. Given a group of subjects assumed to be a random sample from some larger population, the mean of their difference scores should not differ significantly from zero, the value of the population mean which is assumed under the null hypothesis. If the population standard deviation were known, so that the standard deviation of the distribution of sample means could be obtained by $\sigma_{\bar{X}} = \sigma / \sqrt{n}$, the null hypothesis could be evaluated by determining the probability of obtaining a z this large or larger by chance (where $z = (X - 0)/\sigma_X$), simply by using the tables of the normal curve. This procedure, of course, would only be appropriate with samples large enough to warrant the assumption that the distribution of sample means approached normality (say, $n =$ at least 30).

When the experiment involves two different groups of subjects, each of which may reasonably be taken to be a random sample from the same population (an assumption which is usually implemented by random assignment of subjects into the two conditions), under the null hypothesis the difference between their sample means should not differ from zero. In order to construct the appropriate z score for entering the normal distribution in this case, the standard deviation of the difference between two means must be divided into $(\overline{X}_1 - \overline{X}_2) - 0$, since $(\overline{X}_1 - \overline{X}_2)$ is the equivalent of X in the original example of this method. Assuming that the obtained scores in the two samples are uncorrelated (i.e., assuming that the two samples are actually independent), the standard deviation of the difference between two means may easily be obtained from the two separate standard deviations, since

$$\sigma_{\bar{x}_1 - \bar{x}_2} = \sqrt{\sigma_{\bar{x}_1}{}^2 + \sigma_{\bar{x}_2}{}^2}.$$

Since the null hypothesis assumes that the two groups came from the same population, only a single population standard deviation must be known. Again, the central limit theorem requires reasonably large samples for this use of the normal curve.

How Is σ Estimated?

It has been convenient for expository purposes to set aside the problem of how the population standard deviation, σ, is determined in actual practice. In those applications of statistical inference with which psychological experimenters are principally concerned it is necessary to *estimate* the population variance (and standard deviation) from the sample variance, s^2.[1] As a result of the fact that s^2 is employed in place of σ^2, the t distribution must be used instead of the normal distribution for testing hypothesis about sample means. However, aside from the fact that t is used rather than z, the principles of statistical inference outlined briefly above remain essentially the same.

[1] $s^2 = \dfrac{\Sigma(X - \overline{X})^2}{n - 1}$, while $\sigma^2 = \dfrac{\Sigma(X - \mu)^2}{n}$

CONCEPT OF DEGREES OF FREEDOM

In describing how the variance of a sample is used to estimate the population variance, reference will be made to the number of degrees of freedom (*df*) associated with each particular estimate. Since this concept plays a role of fundamental importance and is referred to many times, it will be useful to make its meaning as clear as possible.

If we were told that the mean of a particular sample of scores was equal to 50 but that the sample contained only a single case, we would, of course, know for certain what that score was (we start at this admittedly absurd point for illustrative purposes only). On the other hand, if we knew that the mean of a sample was 50 but the sample contained two cases, we would still need to know the value of one of the two scores before we knew all there was to know about the sample. In this case there is one *df*. If the sample had three cases in it and had a mean of 50, we would have to know two of the three scores in order to know all. In this case there are two *df*. The number of *df* in these and similar examples refers to the number of independent pieces of information capable of being learned about the scores in the sample. When we know $n - 1$ scores, in addition to the mean, there is nothing more to learn. Another way of making the same point is to note that one *df* is lost for each restriction placed on the sample's capability of variation. Adding up the scores in a sample, or computing the mean, reduces our uncertainty about what the sample is like.

Another way of viewing the matter of *df* is in regard to the basic meaning of *variance*, itself. The variance of a set of scores equals $\Sigma(X - \overline{X})^2/n$ by definition. However, when we wish to use the information available in a sample of scores as the basis for obtaining an estimate of the *population* variance, we must take cognizance of the fact that the deviations available for our use are taken with reference to the mean of the *sample*, not the mean of the population. To be sure, in this case the mean of the population is truly unknown. What we do know, nevertheless, is that each deviation that we use to compute our estimate is inaccurate exactly to the extent to which the mean of the particular sample differs from the mean of the population. That is, the difference between each score in the sample and the mean of the population can be conceived of as a double-difference, containing two algebraically additive parts.

The first of these two additive parts is the actual difference between the individual score and the mean of the sample. These will differ from score to score, as the scores themselves vary around their mean. The second part of the double-difference between the score and the mean of the population is the difference between the sample mean and the population mean. This component of the double-difference is the same for all of the scores in the sample, fixed by the particular value of the sample mean. It is this constant piece of unavailable information that produces the loss of one *df* in the sample variance as an estimate of the population variance. Because of this lost piece of information, it is necessary to divide the sample sum of squares, $\Sigma(X - X)^2$, by $n - 1$ instead of by n to obtain an *unbiased* estimate of the population variance from the sample. The extent to which the sample variance, computed as $\Sigma(X - \overline{X})^2/n$, is an underestimate of the population variance decreases as sample size increases. The larger the sample, the less the bias, since the ratio of $n/n - 1$ approaches unity as the size of the sample increases. In any case, dividing the sample sum of squares by $n - 1$ eliminates the bias.

ANALYSIS OF VARIANCE

In Chapter 6 it was pointed out that the logical basis upon which a quantitative statement may be made regarding the significance of an observed difference between group means lies in the fact that the average difference between means obtained by random sampling *from the same population* depends upon the size of the samples, on one hand, and the variability of the measures, on the other. The larger the samples, the smaller the average difference expected between their means (obviously, if each sample contained the entire population, their means would be identical); the more variable the measures, the larger the average difference expected between the means of random samples from the same population.

When there are only two means to be compared, the experimenter employs the variations among measures *within* the two conditions to estimate the variability of the measures in the population. From this estimate of the population variance, he computes the estimated value of the standard deviation of a distribution of randomly selected differences between sample means, the standard error. Dividing this standard error into the obtained difference between the means gives him a t ratio, the distribution of which is known for samples with varying numbers of *df*, and the expected value of which under the null hypothesis is zero. He is, thus, able to state the probability of obtaining a t of this size or greater when the two samples of the given size are selected randomly from the same population, i.e., when the null hypothesis is true. If the probability he obtains is very small, he may choose to reject his assumed null hypothesis (this is, after all, the purpose for which it was assumed) and conclude that the obtained difference was probably not due to chance but to something that he did during the experiment.

When more than two conditions (and means) are to be compared, the t test cannot be used except in special cases as indicated in Chapter 8. It turns out, fortunately, that a more general procedure is available, which in the special case of only two groups results in a ratio equal to t^2. This more general quantity is known as the F ratio, after its creator, R. A. Fisher. Its logic is similar, but not exactly identical, to that of the t ratio.

Since the means of the various groups (or under the various conditions) are assumed under the null hypothesis to have been obtained by random sampling from the same population, an unbiased estimate of the variance of that population may be obtained from the variability among the sample means, since $s^2 = ns^2_{\bar{x}}$. At the same time, the variability of the measures within each of the conditions may be used to generate another estimate of the population variance, as in the fashion of the t test. Where in the t test two separate groups provide a combined estimate of the population variance, in the F ratio the several estimates from within the different groups are pooled into a single estimate. The F ratio is formed from the two separate estimates of the population variance just described, one obtained from the variations among the several groups' means (placed in the numerator) and the other obtained from the pooled within-groups variations (placed in the denominator).

Homogeneity of Variance

It is assumed that these separate estimates of the population variance are *independent* of each other, so that the distribution of F may be employed to determine probabilities. Independence here simply means that the two sources of variability are both assumed to be random, under the null hypothesis, and, therefore, uncorrelated with each other. As was mentioned in the discussion of transformation of the dependent variable, in Chapter 3, this assumption is sometimes violated for one reason or another. The resulting heterogeneity of variance becomes a potential problem only when the different groups in the experiment have widely different numbers of subjects. The consequences of a slight degree of heterogeneity of variance are negligible.

The distribution of F ratios for various numbers of *df* in the numerator and denominator is known and its probability tables may be found in the Appendix. Under the null hypothesis that nothing but random sampling is responsible for any differences among the means under the various conditions, the expected value of F is 1.0 (i.e., the two estimates of the population variance should not differ from each other, since they are both estimates of the same thing). If the null hypothesis is false and there is a real basis for the means of the different groups to differ from each other, the ratio becomes larger than 1.0. Just as in the case of the t test, the probability of their occurrence by chance can be stated only by assuming a null hypothesis which states that chance is responsible for the obtained differences.

Interpretation of a Significant *F*

In an experiment involving more than two experimental conditions or groups, a statistically significant F ratio formed from two independent estimates of the population variance, obtained from the variations among the groups' means and from within the various groups, justifies statements only regarding the *overall* set of differences among all of the groups involved, taken together. In other words, this F ratio does not necessarily indicate that the mean of any particular group was significantly different from the mean of another. It indicates merely that the *entire set* of group means differ among themselves to an extent which justifies the rejection of a chance hypothesis. Significance tests appropriate for differences between selected *pairs* of means must take into consideration how many such pairs could be formed from all the means involved. Obviously, the greater the number of separate means involved, the more likely it is that selected pairs may differ by chance alone. This matter is considered in greater detail in Chapter 8.

The notion of a double-difference, employed above in connection with the concept of *df*, is useful in explaining the basic logic of analysis of variance as well. When k different groups of randomly assigned subjects receive different experimental treatments and each group contains n subjects, the *total* variability of the $kn = N$ observations has $kn - 1$ *df*, for reasons already stated. In other words, the total variance (i.e., deviations taken from the overall mean) has $kn - 1$ *df*. When one thinks of the capability of variation of one of the scores of any one of the groups, it can be recognized that this score may differ from the mean within its own group and, in addition, the mean of its group may differ from the overall mean. The aim of analysis of variance is to segregate these portions of the total variability so that separate variance estimates may be determined for each "source." If the hypothetical score in question is referred to as X_{ij} (the $_i$th score in the $_j$th group), the deviation between it and the total mean, thus, can be seen to contain two additive parts:

$$X_{ij} - \overline{X}_t = (X_{ij} - \overline{X}_j) + (\overline{X}_j - \overline{X}_t),$$

where X_t equals the overall mean and X_j equals the mean of the $_j$th group. The total sum of squares, obtained by squaring and summing the deviations for all of the *kn* scores, likewise, contains two additive parts:

$$\sum_1^k \sum_1^n (X_{ij} - \overline{X}_t)^2 = \sum_1^k \sum_1^n (X_{ij} - \overline{X}_j)^2 + n \sum_1^k (\overline{X}_j - \overline{X}_t)^2$$

$$(\text{Total}) \quad = \quad (\text{Within}) \quad + \quad (\text{Between})$$

In the same way that the total sum of squares is partitioned into its additive components, the total *df* may be similarly partitioned:

$$kn - 1 = k(n-1) + k - 1$$

$$(\text{Total}) = (\text{Within}) + (\text{Between}).$$

When each of the separate sums of squares is divided by its appropriate *df*, the two variance estimates necessary to form an *F* ratio are obtained:

$$F = \frac{\text{Between variance}}{\text{Within variance}} = \frac{n \sum_1^k (\overline{X}_j - \overline{X}_t)^2 / k - 1}{\sum_1^k \sum_1^n (X_{ij} - \overline{X}_j)^2 / k(n-1)}$$

Estimate of Error

Although it is true that both of the variances employed in the *F* ratio just described purport to estimate the same population variance, assuming that the null hypothesis is true, the denominator term is generally distinguished from the numerator by being referred to as the "error term." This distinction, which is the statistical way of noting that the denominator term results from random replication (in the sense in which that term was used in Chapter 6), becomes somewhat more meaningful when one considers what might happen to the two variance estimates in the *F* ratio if the null hypothesis were actually false.

If the independent variable upon the basis of which the groups were differentiated results in different levels of performance, the variance estimate based upon differences among the sample means will no longer accurately indicate the "population" value. Since the different treatments have different effects in this instance, the means would not really be "randomly sampled" from the same population; they would, in fact, have come from different populations (i.e., populations whose means differ from one another). One would expect, in this case, that these means would differ from each other *more* than would be expected had they all come randomly from the same population. The variance estimate in the denominator, on the other hand, should not be influenced by this possibility of real differences among the sample means.

The denominator variance estimate, the "error term," represents a pooling of the variability from *within* the separate groups. Within each group all of the subjects are treated in the same way; they should differ from each other only due to random sampling, i.e., chance. Even though the means of the groups may differ due to the treatments, the deviations which lead to the within variance are obtained with reference to the mean of the particular group within which each score is located. In other words, the separate within-group variances are all assumed to arise from the same population whether or not the null hypothesis is true. Thus, the within variance is referred to as the error term, since *it* reflects the basic random variations in the dependent variable that exist regardless of which experimental treatment a group receives.

In the hypothetical experiment under discussion, it should now be clear that the within variance is the direct result of the manner in which the experimenter has employed the principle of randomization. Subjects were taken randomly from some population (or were assumed to have been obtained randomly); these subjects were next assigned at random to the different experimental groups. Thus, the variability of subjects who are treated similarly (i.e., within each group), should simply be a reflection of these randomizing operations and should provide the experimenter with an appropriate error term for evaluating the effect of his independent variable (i.e., the between variance). These same points have been made previously in relation to the t test (Chapter 6).

Even though the principal reason for using analysis of variance is to generate variance estimates for testing hypotheses regarding differences among sample means, it cannot be overemphasized that the device itself is nothing more than a fantastically flexible set of computational rules for partitioning variability. As will become clear in the remainder of this chapter and the chapters that follow, the same logic and the same simple computational procedures apply over and over again. It is of considerable importance, therefore, that the logic and the basic computational rules be well understood.

COMPUTATION OF SUMS OF SQUARES

The total sum of squared deviations, SS_t, is computed in all analyses of variance directly from the individual scores. In the situation described above (k groups, each containing n subjects),

$$SS_t = \sum_1^k \sum_1^n (X_{ij} - \overline{X}_t)^2.$$

This expression can be shown to be identical to

$$\sum_1^k \sum_1^n X_{ij}^2 - \frac{\left(\sum_1^k \sum_1^n X_{ij}\right)^2}{kn},$$

a formula which is most commonly used for actual computations, since it involves nothing but "raw scores." The term following the minus sign is referred to as the "correction" term, C'.

The between-groups sum of squares,

$$n \sum_{1}^{k} (\bar{X}_j - \bar{X}_t)^2$$

is computed from the following raw-score formula:

$$SS_b = \frac{\sum_{1}^{k} (\sum_{1}^{n} X)^2}{n} - C'.$$

Basically, this formula calls for squaring the sum of scores in each group separately, adding these squared terms together, dividing this sum by n, the number of raw scores which were added prior to squaring, and subtracting the correction term. If the groups have unequal n's, each squared term must be divided by its own n prior to summing.

The within-groups sum of squares,

$$SS_{wg} = \sum_{1}^{k} \sum_{1}^{n} (X_{ij} - \bar{X}_j)^2$$

may be obtained directly, by computing

$$\sum_{1}^{n} X_i^2 - \frac{(\sum_{1}^{n} X_i)^2}{n}$$

for each group and adding these together, or, indirectly, by subtracting the SS_b from SS_t, recalling that $SS_t = SS_b + SS_{wg}$. It should be understood that SS_{wg} is a pool of the variability of the scores within each group around the mean of that group. Since each group's separate sum of squares contains $n - 1$ *df*, the pooled within-sum of squares contains $k(n - 1)$ *df*, $n - 1$ for each of the k groups.

TWO-WAY CLASSIFICATIONS

The introduction to analysis of variance presented above dealt with an experimental arrangement commonly known as a simple, or one-way *random groups design*. This label identifies both the fact that a single independent variable defines the grouping and that different groups of randomly assigned subjects are run under the different experimental conditions. A common elaboration of this approach to experimental research involves a second independent variable in an *orthogonal* arrangement with the first. The word "orthogonal" means that *each level of each* of the independent variables is combined with each of the others to form a separate treatment-group.[2] For example, if one independent variable involved the degree of food–deprivation administered to the subjects (say, 12 or 24 hours), while the other independent variable involved the amount of food reward they were given following the performance of some response (say, 1, 2, or 3 food pellets), an orthogonal arrangement would produce six groups of subjects, the first being deprived of food for 12 hours and receiving 1 pellet reward, the second deprived for 12 hours and receiving 2 pellets reward, the third deprived for 12 hours and receiving 3 pellets, the fourth deprived for 24 hours and receiving 1 pellet, the fifth deprived for 24 hours and receiving 2 pellets, and the sixth deprived for 24 hours and receiving 3 pellets.

The subjects are randomly assigned to the different conditions, as in the simpler experiment considered previously. Again, the variability of subjects treated similarly provides the estimate of error, except that in the present case each group of subjects represents a particular combination of levels of the two independent variables. With two independent variables in the situation, it is no longer appropriate to refer to a single null hypothesis. In fact, with a double-classification system, differences among scores may be associated with either of the two separate classifications and, in addition, to the unique *combinations* of classifications. The best way to explicate this statement is to consider the content of Table 7.1, based on the hypothetical experiment mentioned at the beginning of this section.

With respect to the possible effect of hours of deprivation on the scores in this table, a null hypothesis regarding the two marginal *row* means must be tested. Regarding the possible effect of amount of reward on the scores, a null hypothesis must be tested with respect to the three marginal *column* means. Since it is possible, in addition, for particular *combinations* of hours of deprivation and amount of reward to be associated with higher or lower scores in comparison with other particular combinations of the two variables, still another null hypothesis must be tested regarding the remaining variations among the *cell* means (each row–column intersection defines a *cell*), the last effect being referred to as the *interaction* of the first two variables.

A particular score in Table 7.1 may differ from the overall mean to the extent that the mean of its own cell differs from the overall mean and to the extent that it differs from the mean of its own cell. The latter of these two additive components of variability is analogous to the within-group component of the more simple analysis presented previously. It provides the estimate of error, as before. The former, however, requires

[2] When both of the variables in such an arrangement are *genuine* independent variables (rather than control variables), the design is referred to as a *factorial* design. These designs are discussed in Chapter 8.

TABLE 7.1

Data-analysis table for hypothetical experiment on deprivation and reward.

		Amount of Reward: 1	2	3	Marginal Total Means
Hours of Deprivation	24	$\bar{X}_{24-1}$ n subjects	$\bar{X}_{24-2}$ n subjects	$\bar{X}_{24-3}$ n subjects	$\bar{X}_{24}$ $(3)(n)$ subjects
	12	$\bar{X}_{12-1}$ n subjects	$\bar{X}_{12-2}$ n subjects	$\bar{X}_{12-3}$ n subjects	$\bar{X}_{12}$ $(3)(n)$ subjects
Marginal Total Means		$\bar{X}_1$ $(2)(n)$ subjects	$\bar{X}_2$ $(2)(n)$ subjects	$\bar{X}_3$ $(2)(n)$ subjects	$\bar{X}_t$ $(2)(3)(n)$ subjects

additional partitioning. The mean of a particular cell may differ from the overall mean because the cell is in a *row* whose marginal mean differs from the overall mean. A particular cell mean also may differ from the overall mean because the cell is in a *column* whose marginal mean differs from the overall mean. Furthermore, even if its row and column marginal means are both equal to the overall mean, the particular cell mean may still differ from the overall mean. The number of *df* for the differences among the cell means, following the logic presented earlier, equals the number of cells minus one, $rc - 1$ (r = number of rows and c = number of columns). These $rc - 1$ *df*, and the sum of squares associated with them, may be partitioned into the three additive components alluded to above, with the *df* for the differences among the marginal row means being equal to the number of rows minus one, $r - 1$, the *df* for the differences among the column marginal means equal to the number of columns minus one, $c - 1$, and the *df* for the remaining variations among the cell means equal to the product of $(r - 1)(c - 1)$. In the hypothetical experiment on deprivation and reward, there are six cells (2 deprivations × 3 rewards), with 5 *df*, and these 5 *df* contain 1 *df* for the row effect (deprivation), 2 *df* for the column effect (reward), and 2 *df* for their interaction.

The actual computation of the sums of squares needed for the analysis of variance of the 2 x 3 x n scores in Table 7.1 involves only a simple extension of the principles of the previous one-way analysis. The

$$SS_t = \sum^{r}\sum^{c}\sum^{n} X_{ijk}^2 - \frac{(\sum^{r}\sum^{c}\sum^{n} X_{ijk})^2}{rcn}$$

(note that X_{ijk} refers to the $_i$th score in the cell defined by the intersection of the $_j$th row and the $_k$th column). A SS_{cells} may be computed much as the SS_b in the simpler case, by squaring the separate sum of scores for each cell, dividing the sum of these squared quantities by the number of scores in each cell, n, and subtracting C' from this quotient. Subtraction of the SS_{cells} from $SS_{\underline{t}}$ gives $SS_{\underline{w}/\text{cells}}$ as a remainder. The SS_{cells} contains SS_r, SS_c, and $SS_{r \times c}$. SS_r is obtained by squaring the marginal sum of scores in each row, dividing the sum of these two ($r = 2$) squared quantities by the total number of scores contained in each sum (c columns x n), and subtracting C'. SS_c is obtained by squaring the marginal sum of scores in each column, dividing the sum of these three ($c = 3$) squared quantities by the total number of scores contained in each sum (r rows x n), and subtracting C'. The $SS_{r \times c}$ is obtained as a remainder by subtracting both SS_r and SS_c from SS_{cells}. As in the case of the one-way analysis, each sum of squares is divided by its appropriate *df* to obtain a variance estimate for forming F ratios. In the present example, the within-cell variance estimate serves as an error term for testing the effects of rows, columns, and the row x column interaction. As will be explained below, in certain situations it is necessary to employ the interaction variance estimate as an error term for testing one or both of the other effects.

MATHEMATICAL MODELS FOR ANALYSIS OF VARIANCE

In the two-way experiment considered above, the levels of both of the independent variables were determined *arbitrarily* by the experimenter, to suit his own scientific needs. That is, his special experimental interests determined that he would use the particular amounts of deprivation and reward selected. This method of determining the levels of the independent variables is different, for example, from the alternative procedure of *randomly* choosing a small number of levels of an independent variable from the entire population of possible levels, a procedure which may sometimes be appropriate.

Linear Model for Analysis of Variance

When the experimenter fixes the levels of his independent variables arbitrarily rather than selecting them randomly, the appropriateness of the within-cells variance estimate as an error term may be comprehended in relation to the mathematical model underlying this type of analysis of variance. This model is referred to as the *linear*[3] or the *fixed constants* model. The following equation summarizes the basic assumptions of the linear model:

$$X_{ijk} = \mu + R_j + C_k + I_{jk} + e_{ijk}$$

with μ = hypothetical true score, R_j = row effect, C_k = column effect, I_{jk} = interaction effect, e_{ijk} = error. The sums of the row, column, and interaction effects are each separately assumed to be zero and only the error component is assumed to be normally distributed. Table 7.2 shows the expected[4] values of the variance estimates under the null hypothesis (H_0) and under an alternative hypothesis of real effects (H_1). As can be seen from the expressions in Table 7.2, the within-cells variance pro-

TABLE 7.2

Expected values of variance estimates for the linear model.

Source	Expected values of variance estimate	
	Under H_0	Under H_1
Rows	σ_e^2	$\sigma_e^2 + nc\sigma_R^2$
Columns	σ_e^2	$\sigma_e^2 + nr\sigma_C^2$
Interaction	σ_e^2	$\sigma_e^2 + n\sigma_I^2$
Error	σ_e^2	σ_e^2

vides the appropriate denominator of F ratios for testing the effects of rows, columns, or interaction, since it is the only denominator which would produce departure from an expected $F = 1.0$, when each particular H_0 is false.

The linear model assumes the additivity of the row, column, interaction, and error components. Error is assumed to be normally distributed and randomly sampled. In addition, homogeneity of the separate within-cell components of error variance is assumed.

It should be clear that the foregoing description of the assumptions

[3] Sometimes "additive."

[4] Average value over many samples.

of the linear model for analysis of variance is merely a more mathematical version of considerations previously dealt with solely at a verbal level. The usefulness of a more mathematical presentation will become more apparent later in this section, when more complicated models are presented.

Components of Variance Model for Analysis of Variance

Although experiments involving independent variables whose levels are randomly selected from populations of levels are seldom, if ever conducted in pure form in psychology, it will be useful to consider their underlying mathematical assumptions prior to introducing more frequently used complexities. The underlying model is referred to as the *components of variance* or the *random* model. A hypothetical example of an experiment [5] of this type, involving two independent variables with randomly sampled levels, might take the following form.

Suppose the investigator wished to study the relationship between psychotherapeutic outcome (degree of improvement), on one hand, and both psychodiagnosis and type of psychotherapy, on the other. He is not interested in particular diagnoses or types of therapy, but with the general notion that different types of therapy may have different outcomes depending upon diagnosis. He is also interested in the overall effects of diagnosis and therapy. In other words, his greatest interest is in the possibility of an interaction between diagnoses and therapies, regardless of the particularities of this effect, and he has some interest in the separate variables as well. Because there are long lists of both diagnoses and types of therapy, and because he is interested in no particular instances of either, he randomly selects a manageable number of diagnoses (say about five or six) and an equivalent number of types of psychotherapy. The "experiment" is conducted by administering the selected types of therapy to individuals representing the selected diagnoses, with two or more individuals of each diagnostic category in each psychotherapy condition.

The following equation summarizes the basic assumptions of the random model:

$$X_{ijk} = \mu + u_j + v_k + w_{jk} + e_{ijk}$$

with μ = true score, u_j = row effect, v_k = column effect, w_{jk} = interaction effect, and e_{ijk} = error. The *expected* values of u, v, and w, are zero, and all of the components are random. Table 7.3 presents the expected values of variance estimates under H_0 and H_1 and indicates that the

[5] In the language used in this book, studies of this type would be quasi-experiments, since manipulated independent variables are not used.

TABLE 7.3

Expected values of variance estimates for the random model.

Source	Expected values of variance estimate	
	Under H_0	Under H_1
Rows	σ_e^2	$\sigma_e^2 + n\sigma_W^2 + nc\sigma_R^2$
Columns	σ_e^2	$\sigma_e^2 + n\sigma_W^2 + nc\sigma_C^2$
Interaction	σ_e^2	$\sigma_e^2 + n\sigma_W^2$
Error	σ_e^2	σ_e^2

within-cells variance is *not* the appropriate source of error for testing either the row or column effects, except under very special circumstances.

Only the interaction is normally tested with the within-cells error estimate in this type of study. Because the expected values of the variance estimates for rows and columns each contain a component arising from the randomly sampled interaction, along with error and a component from the effect of the variable, itself, the interaction variance estimate provides the appropriate error term for testing rows and columns.

Under the random model for analysis of variance, the F test for interaction is the same as in the linear model. But the test for rows and columns is complicated by the fact that the expected values of their variance estimates are each contaminated by a component due to interaction. This contamination arises from the fact that random selection of the levels of either of the independent variables automatically results in random generation of values of the interaction of it with the other variable, even if the other variable is of the arbitrary rather than the random type. In other words, at the same time that, say, *paranoid schizophrenia* is selected at random to be one of the levels of the diagnosis variable, any unique vulnerability or invulnerability of this particular disorder to treatment by one of the types of therapy is also automatically sampled, thus contaminating the variance estimate of the other variable. As will be seen later, this contamination has consequences of importance in analyzing experimental data.

Mixed Model for Analysis of Variance

Often, in psychological experimentation the levels of one of the independent variables in multivariable studies are randomly sampled while the levels of the others are arbitrarily fixed by the experimenter. Studies of this type are considered in detail in Chapter 10. The randomly selected variable in question in all of these cases is the experimental *subject;* that is, a random sample of subjects is taken to represent some population for which the results of the study have relevance. Each subject performs two or more times under all of the levels of one or more of the fixed independent variables. The underlying mathematical model for the analysis of variance is referred to as the *mixed* model, since one of the variables is of the random type and one or more are of the fixed type. Table 7.4 presents the expected values of the variance estimates under H_0 and H_1 for this situation.

TABLE 7.4

Expected values of variance estimates for the mixed model.

	Expected values of variance estimate	
Source	**Under H_0**	**Under H_1**
Rows (Fixed)	σ_e^2	$\sigma_e^2 + n\sigma_W^2 + nc\sigma_R^2$
Columns (Random)	σ_e^2	$\sigma_e^2 + nr\sigma_C^2$
Interaction	σ_e^2	$\sigma_e^2 + n\sigma_W^2$
Error	σ_e^2	σ_e^2

Reference to Table 7.4 will reveal that the expected value of the variance estimate for the row effect (the variable whose levels were determined arbitrarily by the experimenter) contains a component resulting from the randomly sampled interaction, $n\,\sigma_W^2$, along with $nc\,\sigma_R^2$ and σ_e^2. For this reason, the appropriate error estimate for testing the null hypothesis regarding the row effect (the fixed effect) is the interaction variance estimate, since this F ratio will differ from 1.0 only to the extent that a real source of $nc\,\sigma_R^2$ is present. This example illustrates a general rule of thumb regarding the formation of F ratios with appropriate denominators, namely, that the denominator term should contain all of the components of the numerator with the exception of the variance due to the independent variable itself (cf. Schultz, 1958). A significant F ratio in this case would be interpreted to mean that the variable in question had an effect over and above any possible interaction in which it was involved. If the within-cells error variance estimate were used in the denominator of the test of the row effect, a significant F could mean either that differences were associated with the variable itself, or its interaction with the random variable, or both. Psychological experimenters are generally more interested in the former of these tests, since the more general test is not unambiguously interpretable. However, in certain types of practical situations it may be extremely useful for the experimenter to learn that the variable has some kind of effect as opposed to none whatever. A more detailed discussion of this matter may be found in Cox (1958), including an informative experimental example.

The column effect is uncontaminated by the interaction component, since the levels of the row variable were not randomly sampled. Thus, the column effect is tested with the within-cells error variance estimate as in the linear model. Similarly, the interaction effect, itself, is tested with the within-cells error variance estimate. These and related points are discussed further in Chapter 10, in the section dealing with the treatments x subjects design.

STATISTICAL ADJUSTMENT OF THE DEPENDENT MEASURES

Reducing Error Variance Statistically

The principal mission of experimental design is to permit appropriate statistical tests to be made in relation to the experimental hypotheses, maximizing sensitivity while minimizing experimental effort (e.g., keeping the size of *n* as small as possible). In the foregoing discussion of statistical inference, it has more or less been assumed that whatever the error variance happens to be, it is. The problem dealt with so far has been that of identifying and computing the appropriate source of error variance for testing the significance of the various experimental effects, rather than that of attempting to increase sensitivity by reducing the error. Chapters 9, 10, and 11 of this book describe a variety of *experimental* methods for reducing error. Because it was thought desirable to concentrate completely *statistical* considerations within a single chapter, error-reduction by statistical adjustments of the dependent variable is dealt with in this section.

The most significant source of error variance in psychological experimentation arises from the fact that individual subjects differ widely in the various abilities that determine performance on the dependent variable. Generally speaking, this variability exists both before and after the various experimental treatments are administered. Because relatively reliable characteristics of behavior are responsible for this "noise" background against which the significance of experimental effects must be assessed, adjustments can often be made in the dependent variable measures which have the effect of more or less "taking out" the differences in question. In Chapter 3, for example, it was pointed out that the dependent variable may sometimes be a difference-score (or a change-score), with a pre-experimental measure being subtracted from the observed dependent measure, or vice versa.

Adjustment by Using Difference-Scores

The appropriate error variance for testing hypotheses regarding mean differences among adjusted difference-scores, obviously, must be found in the variability in the derived measures rather than variability in the original dependent measures. Since each dependent measure may be conceptualized as containing two additive components, one due to the pre-experimental individual difference in ability and the other due to the effect of the experimental treatment (plus random error in the latter), elimination of a highly variable component from each of the scores by subtraction should have the desired effect of reducing the error variance. In general, adjustment of the dependent variable by the difference-score method assumes that the pre-experimental measure is linearly correlated with the unadjusted dependent measure and that this correlation is sufficiently great to result in actual reduction in variability. Most writers on this topic agree that a correlation of 0.5 is an effective minimum standard. It must be emphasized that the pre-experimental measure is assumed to be uncorrelated with the independent variable, since it would be totally devastating if the adjustment tended to have different effects in the different experimental conditions. Another way to put this last point is to note that the correlations between the pre-experimental measures and the unadjusted dependent measures in the separate experimental groups should all be estimates of the same population correlation. As is noted later in connection with the discussion of analysis of covariance, this is known as the assumption of homogeneity of regression.

Adjustment by Using Ratio-Scores

Experimenters occasionally employ the adjustment device of expressing the observed dependent measure as a proportion of percentage of some pre-experimental measure. Viewed as a potential alternative to the difference-score method, these ratio-scores are seldom to be preferred. The principal reason for this statement is the fact that even the most sophisticated psychological research almost never involves measurement scales possessing more than equal-interval properties. Since equal-interval scales involve arbitrary and often unknown true zero points, the formation of ratios by dividing one such score into another is apt to increase variability as often as it reduces it, as well as possibly generating quite misleading relative differences. For example, suppose Subject No. 1

had a pre-experimental score of 5 and an unadjusted dependent variable score of 20, while Subject No. 2 had a pre-experimental score of 10 and an unadjusted dependent variable score of 20. Subject No. 1's ratio score is twice as great as that of Subject No. 2 (20/5 compared to 20/10). Now, suppose in addition that the arbitrary zero point of this equal-interval scale is actually 100 points above the "true" zero. Subject No. 1's "true" ratio score is actually 120/105, or 8/7, and Subject No. 2's "true" ratio score is actually 120/110, or 12/11. The difference between these "true" ratios is far smaller than the difference between the ratios actually computed. Had difference-scores been used, on the other hand, the relationship between the scores of the two subjects would be the same regardless of the location of the arbitrary zero point.

As is noted in Chapter 8, when pre-experimental measures are available at a time which permits the use of an *experimental* arrangement for the reduction of error, it is usually preferable to employ such an arrangement. Often however, such measures are obtained during the same experimental session in which the unadjusted dependent variable is also observed. Under conditions such as these, naturally, it is impossible to utilize the concurrently collected information as a basis for assigning subjects into experimental conditions. If error-reduction is to be achieved in these cases, statistical adjustments must be made "after the fact." The most direct approach is the difference-score method presented above.

Analysis of Covariance

A more elaborate approach known as *analysis of covariance* may also sometimes be employed, utilizing the logic of correlational prediction as a device for segregating that portion of the total variance in the dependent variable which is predicted by the correlation from that portion which is not. The basic building block in the logic of analysis of covariance is the same as that previously mentioned in connection with the difference-score method, namely, that each individual unadjusted dependent measure may be conceptualized as containing two additive components. One component is predicted from the correlation between the pre-experimental variable (the "covariate") and the unadjusted dependent variable, and the other component is unpredicted. The appropriate F ratio is formed by dividing the unpredicted variance within groups into the unpredicted variance between groups, i.e., the equivalent of an ordinary analysis of variance is conducted using only that portion of the total variability in the dependent variable that is unpredicted by the correlation. The interpretation of a significant F ratio is much the same as in ordinary one-way analyses, except that the experimenter may focus attention on the *adjusted* dependent variable.

Example of Analysis of Covariance

To illustrate the statistical logic as well as the actual computational procedures for conducting an analysis of covariance, an experimental example will be useful (Sandler, 1971). Three groups of randomly assigned rats are given different amounts of postcriterion training in an avoidance learning situation (amounts of avoidance training, A, has $a = 3$ levels). Each group contains n animals. Following avoidance training, all of the subjects are tested in a punished-extinction task, in which each previously effective avoidance response results in immediate punishment. The dependent variable (in this example, Y) is the number of trials taken to reach the extinction criterion. The experimenter has reason to believe that, because different animals required different numbers of avoidance training trials *prior* to achieving the avoidance criterion (and, thus, prior to the introduction of the independent variable), the number of precriterion training trials would be positively correlated with resistance to extinction. To remove this source of variability from the dependent measures, an analysis of covariance is conducted using the precriterion trials measure as the covariate (X).

The analysis of covariance is concerned with that part of the total variability in Y which is unpredicted by the correlation between X and Y, number of precriterion avoidance trials and number of trials to extinction. From correlation theory we know that $(\Sigma xy)^2/\Sigma x^2$ equals that portion of the total sum of squares in the dependent measures, $\Sigma(Y - \overline{Y}_t)^2$, that is predicted by the correlation. In this formula, $x = X - \overline{X}_t$ and $y = Y - \overline{Y}_t$. A raw-score formula for computing $(\Sigma xy)^2/\Sigma x^2$ is:

$$\left(\sum_1^a\sum_1^n XY - \frac{\left(\sum_1^a\sum_1^n X\right)\left(\sum_1^a\sum_1^n Y\right)}{an}\right)^2 \Bigg/ \left(\sum_1^a\sum_1^n X^2 - \frac{\left(\sum_1^a\sum_1^n X\right)^2}{an}\right) = SS_P.$$

When SS_P is subtracted from SS_T, the total sum of squares in the dependent variable, the total unpredicted sum of squares, SS_{Un}, is obtained.

Contained in SS_{Un} is a portion of unpredicted variation within the different experimental conditions $SS_{Un}/_W$ and a portion between experimental conditions, $SS_{Un}/_B$. The former provides an estimate of error for the F test, while the latter provides a measure of the adjusted (or "apparent") treatment effect. $SS_{Un}/_W$ is obtained as above by subtracting $(\Sigma xy)^2/\Sigma x^2$ from the total within-group sum of squares, SS_W, except in this instance the X and Y deviations are taken from the *group* X and Y means rather than from the *total* means. SS_W is computed as in any one-way analysis of variance and the within-groups $(\Sigma xy)^2/\Sigma x^2 = (SS_P/_W)$ is subtracted from it to give $SS_{Un}/_W$. A raw-score formula for computing $SS_P/_W$ is:

$$\left(\sum_1^a\sum_1^n XY - \frac{\sum_1^a(\sum_1^n X)(\sum_1^n Y)}{n}\right)^2 \Bigg/ \left(\sum_1^a\sum_1^n X^2 - \frac{\sum_1^a(\sum_1^n X)^2}{n}\right) = SS_{P/W}.$$

Subtracting the $SS_{Un/W}$ from SS_{Un} gives the adjusted treatment sum of squares, with $a - 1$ *df*. $SS_{Un/W}$, with $a\,n - a - 1$ *df*, provides the appropriate error estimate.

The computational procedure for analysis of covariance, thus, may be summarized as follows:

1. Compute SS_T conventionally.
2. Compute the total predicted sum of squares, $(\Sigma xy)^2 / \Sigma x^2 = SS_P$.
3. Subtract SS_P from SS_T to get the total unpredicted sum of squares, SS_{Un}.
4. Compute SS_W conventionally.
5. Compute the within-groups predicted sum of squares, $(\Sigma xy)^2 / \Sigma x^2 = SS_{P/W}$.
6. Subtract $SS_{P/W}$ from SS_W to get $SS_{Un/W}$. This is the error term for the F test.
7. Subtract $SS_{Un/W}$ from SS_{Un} to get the adjusted treatments sum of squares for the F test.

Assumptions Underlying Analysis of Covariance

The assumptions underlying analysis of covariance are more stringent than those involved in simple analysis of variance. This is probably the main reason that experimental methods of error-reduction are generally preferable whenever possible. Analysis of covariance requires not only additivity, homogeneity of variance, and normality of error, the three main assumptions of analysis of variance, but also the additional assumptions of homoscedasticity (homogeneity of the Y variances across values of X) and homogeneity of regressions among the experimental conditions. The latter of these additional assumptions is also, it will be recalled, implicitly required when difference-scores are used to adjust the dependent measures. Analysis of covariance would be the most effective method for error-reduction of any method available were it possible to ensure in advance that its required assumptions could be perfectly met. Since this is a practical impossibility, the use of analysis of covariance in psychological experimentation is limited to those situations in which pre-experimental measures cannot be obtained at a time early enough to permit experimental methods of error-reduction to be used.

Tests are available in standard statistical texts for evaluating the assumption that the separate within-group regressions are homogeneous and for establishing that the overall regression is significant. The latter requirement (i.e., significant regression overall) is seldom worth testing, since analysis of covariance is not an effective procedure except when $r > 0.5$ anyway. When $r > 0.5$, the regression would almost always be significant, except in experiments with very small *n*'s. The analysis of covariance method is also applicable in factorial experiments and even in experiments involving error-reduction by experimental means.

ANCILLARY ROLE OF STATISTICS

In a chapter devoted primarily to statistical considerations, it may seem paradoxical to conclude with a section emphasizing the adjunctive, secondary status of such methods. The importance of statistics notwithstanding, it is of far greater importance to establish and maintain the perspective of the empirical scientist for whom statistics is a tool and to avoid a not-infrequently-observed tendency to permit the statistical cart to get before the empirical horse. The scientist conducts experiments for the purpose of obtaining empirical answers to his questions about the nature of objects, organisms, and events in the world about him. When he comprehends and is in control of most of the variables that influence a particular class of behavior, he can conduct experiments whose results are interpretable with little or no statistical analysis at all. When the experimenter must employ statistical techniques for establishing the significance level of the obtained results, he employs these devices to suit *his* purposes and to assist him in communicating with his scientific colleagues.

Regardless of what the stated statistical probabilities may be, the psychological conclusions arrived at by the experimenter may still either be accepted or questioned by his fellow scientists. The probabilities associated with his statistical tests are merely one facet of his argument. Whether statistics plays a large or a small part in establishing the credibility of the argument, it remains at all times *ancillary* to the main message. The principal responsibility of the experimenter is to describe *what* was done and *why* it was done, and, of course, what *happened.* No matter how important a statistical test may be in supporting his arguments and conclusions *about* what happened, the actual observations of behavior (whether or not reduced to averages and standard deviations) are *always* more important. Although some readers may find this point so obvious as to be unnecessary, the ease with which one can find published scientific reports whose results sections begin with analyses of variance rather than graphs or tables of response measures (i.e., *results*) testifies that it must still be made.

Tukey (1969) has observed that it would be better to bend the data so that a particular analysis may be done and an important question answered, than to answer some undesired question or no question at all because of rigid acceptance of the rules laid down by mathematical statisticians. As Tukey has so aptly phrased this point, "I know of no thoughtful account of scientific research . . . according to which data analysis should fail to be *flexible,* or should fail to be a *handmaiden* rather than a high priestess" (1969, p. 90, emphasis added). For this reason, this book emphasizes experimental logic rather than mathematical principles in dealing with many matters that are often thought of as being statistical.

A good example of the potential high priestess role played by statistics may be seen in the procedure of selecting a significance level before an experiment is begun and then reporting the results of all statistical tests as "significant" or "not significant," with respect to the a priori chosen value. In such cases, two t ratios are both reported as "significant," even though the tabled probabilities of the two may range from 0.05 to 0.001, or even lower. Whatever the chances may be of obtaining a particular value of t in a particular experiment, they are surely not influenced by the experimenter's a priori decision as to which level to use. Adherence to "rules" of this type justifies the belief expressed by some (e.g., Skinner, 1956) that statistics serve primarily honorific purposes.

Another example may be found in the notion that "directional" hypotheses regarding differences between group means permit "one-tailed" statistical tests to be employed, simply because mathematical statisticians have educated psychologists to the fact that the likelihood of obtaining a difference of a certain size *in a particular direction* (e.g., mean A larger than mean B) by chance is exactly one-half of the likelihood without regard to direction. The logical model upon which tests of the significance of differences between sample means are based implies, however, a nondirectional null hypothesis. The experimenter (*not* the statistician) begins with the assumption that only chance has influenced his observations under the two conditions. The "chance" he is thinking of is random and is equally capable of generating errors in either direction. This is the "chance" whose probability tables he employs when significance levels are determined. There is nothing statistical in his decision to divide the tabled probability value in half; yet he does so because statisticians have told him that one-tailed tests are more powerful than two-tailed tests. As if the empirical science of psychology could be moved *more* by an observation of a difference that could have occurred by chance as often as ten times in a hundred simply because the experimenter has pointed out that it could have occurred *in that direction* by chance only five times in a hundred, rather than by the experimenter "telling it like it is."

If reliance upon mathematical logic is paramount in the experimenter's thinking, his work will be guided more by that logic than by an active empirical sensibility. The author takes rather the opposite position on this question. According to this position, the experimenter should attempt at all times to focus on the observational and analytical psychological purposes of his work and should use statistics only as he sees fit, regardless of the "rules" he may violate. If he succeeds in maintaining an empirical outlook, he will discover that many design and analysis decisions can be made by *him* on scientific grounds rather than by statisticians to whom he might delegate this function.

SUMMARY

The hypothesis that a particular score, X, was obtained by random sampling from a normal distribution with known mean and variance may be tested by converting X to z and employing the tables of the normal distribution to determine the probability of occurrence of a z of that size or larger. The same logic applies to sample *means* except that the z in this case is produced by dividing by the *standard error* of the mean, again illustrating the principle of appropriateness of the error term in hypothesis testing. Differences between pairs of means may be evaluated similarly. Since the form of the *sampling distribution* of means approaches normality as sample size increases (the *central limit theorem*), and since most null hypotheses assume a population mean of zero, this procedure has had wide application.

Because of the necessity of *estimating the population variance,* the t distribution, with appropriate *degrees of freedom,* has replaced the normal distribution in most cases, but the logical principles remain the same. The term, degrees of freedom, refers to the number of independent pieces of information capable of being learned about the sample scores. An *unbiased* estimate of the population variance is obtained by dividing the sample *sum of squared deviations* by the number of degrees of freedom, usually $n - 1$.

Analysis of variance is a procedure for partitioning the total variation (sum of squares) into its additive component parts. When measures are obtained under several different conditions which define the levels of an independent variable, the total variation may be partitioned into a sum of squares due to differences *between conditions* and another due to differences among measures *within conditions.* Variance estimates are obtained from both of these by dividing each by its own degrees of freedom. A test of the null hypothesis that the different group means were obtained randomly from the same population (i.e., that the independent variable had no effect) is made by dividing the *between variance* by the *within variance,* forming an F ratio, and then looking up the probability of this ratio in published tables.

Under the null hypothesis, the two variance estimates are each independent estimates of the same population variance. If F is significant, the experimenter may conclude that the independent variable influenced the dependent measures differentially in the different conditions. The denominator of this F ratio is called the *error term,* since it is based upon random differences among subjects (or observations) treated similarly. The assumption of *homogeneity of variance* (that the several within-condition variance components are the same) is required.

The same logic applies when there are two or more independent variables arranged *orthogonally.* When each of these independent variables is of interest in its own right (rather than being run merely for control purposes), the experiment exemplifies a *factorial design.* Factorial designs are based upon one of three different underlying *mathematical models for analysis of variance,* which determine the appropriate denominators for forming F ratios. When all factors have arbitrarily determined levels, the model is called the *fixed constants* model; when all factors have randomly sampled levels, the model is called the *components of variance model;* and when both types coexist in the same experiment, the model is called a *mixed model.* Most psychological experiments are of the *fixed constants* or *mixed* types, the latter usually occurring when a random sample of subjects constitutes one of the factors in the design.

When it is possible to obtain concurrent or preliminary measures from the experimental subjects which are correlated with the dependent variable but not with the independent variable, *statistical adjustments* of the dependent measures may result in reduction of the error variance and consequent increase in sensitivity. One frequent approach to this situation is to use *difference scores* which remove the variations among measures existing before the independent variable was administered. A more elaborate and effective method, known as *analysis of covariance,* may also be used, provided that its stringent assumptions are met. Basically, this device attempts to evaluate the effect of the independent variable in that portion of the variation among the dependent measures that is unpredicted by the correlation with the concurrent control measure.

It must be emphasized that in spite of its usefulness, statistics at all times remains ancillary to the main *empirical* thrust of experimentation. It must be a handmaiden, used by the empirical scientist for his purposes, rather than a high priestess, dictating what he can and cannot do. Under conditions of sufficient experimental control, in fact, statistics may become altogether unnecessary.

8

Simple and Factorial Experiments

Experiments are often described as controlled observational situations in which "everything is held constant except the variable of interest" (e.g., Underwood, 1957, pp. 35–36). The "everything" referred to in this brief characterization, of course, consists of all of the potential independent variables known or suspected to have an influence on the dependent measures.

As a beginning point, the "hold everything constant, etc." statement is quite satisfactory, since the *essence* of controlled experimentation does lie in the highly regulated circumstances in which relationships are studied. If one wished, for example, to conduct an experiment on the effect of the level of food deprivation upon the speed of performance of an approach response, the amount of food presented to the subject at the termination of the response would have to be *the same* for all of the subjects, in all of the deprivation conditions (as would the amount of time between completion of the response and delivery of the food, etc.).

On the other hand, if the independent variable of interest were the amount of food presented at the completion of the response, the subjects would all have to be run under identical amounts of food deprivation. *Both* variables in this case are known to influence the speed of response, and the effect of one could not be comprehended unambiguously if the other were also permitted to vary. At this most simple level it is quite easy to recognize that the elucidation of the influence of one variable on the dependent variable may require that other relevant variables be held constant.

The problem is often not quite this simple, however. Frequently, the effect of interest involves the complex influence of two or more independent variables, operating simultaneously, on the dependent variable. Instead of being able to study the effect of each independent variable in isolation, while holding the others constant, it frequently becomes necessary to vary two or more independent variables systematically, so that their joint effects, their *interaction*, may be studied.

While it still may be possible logically to conceptualize certain segments of this type of more complex experiment as exemplifying the principle of holding "everything constant," the separate effects of the different independent variables often do not merely "add" to one another to produce the final result but, instead, combine multiplicatively or, more generally, nonlinearly. In such cases, the *combined* influence of the independent variables participating in the interaction may be of substantially greater importance than the separate influences of the component independent variables themselves. For this reason and others it will be fruitful to restate the description of an experiment given above so that it allows for the deliberate manipulation of several independent variables at once, to permit their combined effects to be studied, with "everything" that is *not* systematically varied either held constant, counterbalanced, or randomized.[1]

SINGLE-FACTOR EXPERIMENTS

Suppose that we wish to study the effect of some independent variable, A, upon a quantitatively measurable dependent variable, and that analysis of the problem prior to data-collection indicates that A be administered at a levels. Although it may be both possible and preferable to employ the principle of *blocking* to deal with the problem of subject heterogeneity (either by using matched groups of subjects or by using each subject more than once, as discussed in Chapters 9 and 10, respectively), we will assume further that each level of A is to be administered to a different group of n subjects. With n subjects in each group and a groups, there are a times $n = an$ subjects in all. This type of experiment involves what is referred to as a *random groups* design.

As was noted in Chapter 6, *randomization* provides the physical (i.e., procedural) basis for the null hypothesis. Simply put, the null hypothesis states that the groups' means differ from one another only to the extent that groups of their size, chosen randomly from the same population, may differ on the average as a result of *chance*. Randomization is

[1] While the details of such procedures are beyond the scope of this book, methods are also available for conducting and analyzing experiments involving the *simultaneous* influence of two or more factors on response surfaces (Box, 1954).

the means by which the experimenter assures that the laws of chance are operative in his situation. In the present instance, randomization is implemented simply by assigning subjects at random to the *a* groups in such a way that each subject has an equal chance of being assigned to each group (this is done most conveniently by using a table of random numbers).

Given that the subjects represent a random sample from some specifiable population (an assumption that is seldom technically true but often not deviant enough from the truth to present a serious problem), the variance of that population may be estimated either from the individual differences among the subjects within each group or from the differences among the groups' means. At this point in the conceptual process, the only differences among the subjects within each group and between the groups' means are those which resulted from the randomization procedure.

The dependent variable measures are obtained after the groups have been differentiated by implementation of the independent variable procedures. Each group receives a different "treatment," i.e., is run at a different level of the independent variable. If the null hypothesis were "true," of course, the different treatments should result in no systematic differences in the measured behavior. If the null hypothesis is false and the independent variable *does* affect the behavior under study, the means of the groups will tend to differ from one another. Even if the null hypothesis is false and a systematic difference among the groups' average dependent measures results, however, the measures made on subjects *within the same group* will still differ *only* as a result of random assignment, since within each group the randomly assigned subjects are all treated alike. Thus, the within-group variance, which is an average of the variability of subjects treated similarly, will still provide an accurate estimate of the population variance.

The statistical test of the null hypothesis is made by comparing this estimate of the population variance obtained from subjects treated similarly (i.e., an estimate that *is not* influenced by the independent variable but only reflects random "error") with one based upon the observed differences among the groups' means (i.e., an estimate that *may* be inflated by the effect of the independent variable if the null hypothesis is false). The tabled values of F and t ratios simply indicate how often different amounts of discrepancy between the two population variance estimates can be expected to occur by chance, given the particular *df* of the experiment in question.[2] In a manner of speaking, these tables rep-

[2] Actually, the t ratio is a special variation of F (in fact, the square root of F in the two-group case), in which the numerator contains the mean difference between groups and the denominator contains an estimate of the standard error of the difference between means.

resent a complete *description* of the null hypothesis. The more deviant the two estimates of the population variance are, the less likely it is that the null hypothesis is true and the more likely it is that the independent variable has had an effect on the measured behavior.

Appropriateness of Error Term

In a single-factor experiment such as the one under consideration, the *appropriateness* of the within-groups variance as a standard for evaluation of the differences found among the groups' means (i.e., as an "error" term for testing the null hypothesis) can be comprehended in relation to just *what* the phenomenon is that is under study and in what way replication has been employed. By "the phenomenon" in this case is meant the differences in the observed behavior associated with the different levels of the independent variable. A single replication of this phenomenon would require only a single subject for each of the a levels of the independent variable, A. Since the a subjects in a single replication of the experiment may differ by chance (they *were* assigned to conditions randomly), the within-groups variance provides *exactly* what is needed as an estimate of error variability in this case, an estimate of the variability of subjects that is due only to randomization. The "phenomenon" must involve differences in excess of this variability, since this is the variability present even when the "phenomenon" is altogether nonexistent. Thus, the basic experiment is replicated n times. Since each of the a groups contains $n - 1$ *df* between subjects, the pooled within-groups variance contains $a(n - 1)$ *df* in all. Table 8.1 shows the breakdown of *df* for the appropriate analysis of variance.

TABLE 8.1

Breakdown of *df* for analysis of variance of an experiment employing a single factor, *A*, at *a* levels, with *n* replications *(random groups design)*.

Source	*df*
A	$a - 1$
Error *	$a(n - 1)$

* Rather than using the word that describes the computational source of the estimate of error, in this case, the pooled within-groups variance, the word identifying the *attributed* source, i.e., random error, will be employed except where indicated.

Homogeneity of Variance

The procedure of pooling the separate within-group variability measures into a single estimate of the population variance, the mechanics of which were presented in Chapter 7, takes quite seriously what the null hypothesis specifies. Since the variability within each group is assumed (because of randomization) to reflect the same population variance, it is quite reasonable to determine it by pooling the separate sums of squared deviations and dividing this sum by the pooled *df*. The logic underlying this procedure, however, requires that this implication of the null hypothesis be satisfactorily justifiable empirically. Thus, the assumption of "homogeneity of variance" may be evaluated separately by a statistical test of the hypothesis that the individual within-group variance estimates all estimate the same population value.

When there are only two groups of measures (i.e., only two levels of the independent variable) it is sufficient simply to form an *F* ratio of the two within-group variance estimates, with the larger term arbitrarily placed in the numerator, and to enter the table of *F* using the appropriate *df*. The probability value arrived at in this way must be doubled, since the table is based on a model in which the value in the numerator could have been in the denominator just as often (but was arbitrarily prevented from being there). When there are more than two levels of the independent variable, a more complex procedure may be required to support the assumption of homogeneity of variance (e.g., Bartlett's test, Edwards, 1960). If the two most divergent variance estimates do not differ significantly in an *F* test of the type described above, it is not necessary to undertake the more complex comprehensive test. Even if it is necessary to employ Bartlett's test, it is probably safe to ignore heterogeneity that does not reach the 0.01 level of significance, both because of the demonstrated robustness of the *F* test (as mentioned below) and because Bartlett's test tends to be somewhat hypersensitive.

Since heterogeneity of variance may have been produced by nonindependence of the mean and variance in the several experimental conditions, an appropriate transformation of the dependent variable may have the effect of eliminating or greatly reducing it. That is, if one of the kinds of measurement restrictions discussed in Chapter 3 has had the effect of inducing correlation of the means and variances among the experimental groups, it may be possible to overcome this correlation by an appropriate transformation of the scale. Elimination of the correlation will usually restore homogeneity of variance in these cases.

However, no transformation may be available that can overcome some particular instance of heterogeneity, as for example, when the basic assumption of *additivity* is unjustified. In this case, some subjects may be influenced idiosyncratically by the independent variable (or some of its levels) as compared with some other subjects. If it is possible in advance to ascertain the nature of the subject variable that might be responsible for this type of nonadditivity, of course, this variable may be manipulated deliberately for the purpose of creating subgroups of subjects that are likely separately to be more appropriately homogeneous in their reaction to the independent variable (e.g., males and females). The experiment would then no longer be of the single-factor type, but would become what is referred to below as a *quasi-factorial* experiment. The problem of heterogeneity of variance may sometimes be dealt with effectively in this way.

Robustness of Statistical Tests

In those cases in which nonadditivity resulting in heterogeneity of variance is unforeseeable and, therefore, unavoidable, it may still be possible to conduct standard statistical analyses with relatively little danger of arriving at erroneous conclusions. If the number of observations under each level of the independent variable is constant (or almost so), very little discrepancy is likely to be produced between computed values of t or F and those that would have been obtained under conditions closer to the required assumptions (Boneau, 1960).

This manifestation of the robustness of the t and F tests is, however, dependent upon equal-sized groups. Very wide discrepancies in group size coupled with significant heterogeneity of variance may result in erroneous conclusions, when the larger variances are associated with the smaller n's. Even under these highly unfavorable conditions, if there are but two groups in the experiment an approximate t test can be performed that takes the inequality of n's into account along with the inequality of variances (Edwards, 1960). If the n's are equal but the variances significantly different, this approximate t test is achieved merely by computing t as usual and doubling the obtained probability. Naturally, it is always possible to use nonparametric statistical tests, the assumptions of which are less stringent and more likely to be justified (Siegel, 1956). These tests, however, are often less powerful than their parametric equivalents and, in addition, frequently are based implicitly upon the same underlying assumptions.

Interpretation of Significance

When there are only two groups in the experiment, a significant t or F can be directly interpreted in relation to the observed difference between the groups' means. That is, the test implies the conclusion that the *ordinal* difference (e.g., the experimental group's mean was higher than the control group's mean) was not due to chance.

When more than two groups are involved in the experiment, the significance of the F test (using the variance among the groups' means to estimate the population variance in the numerator and the pooled within-groups variance to estimate the population variance in the denominator) indicates that the differences associated with the independent variable are greater than would be expected to occur by chance (or, more precisely, that differences of such magnitude would occur by chance only very infrequently). The *entire set* of group means is implicated in this conclusion, so that it would be quite inaccurate to use the overall F as a basis for statements regarding differences between any particular pair of means.

In many experiments involving a single, quantitative independent variable, the actual ordering of the groups' means may be so straightforward and simple as to preclude the necessity of any additional tests beyond the overall F test. But, as is discussed in the next section of this chapter, the overall F test may often be only the precursor of a series of subsequent statistical tests, without which no psychologically useful conclusions could be stated.

Statistical Tests Following an Overall F Test

When the various experimental groups have received *qualitatively* different treatments, it is obviously impossible to arrange them in any quantitatively meaningful order in relation to the levels of the independent variable. Postmortem analysis of the differences between pairs of means may be conducted, nevertheless, to permit separate statements to be made about the significance of the difference between each pair. It is inappropriate simply to choose the larger observed differences for evaluation by means of t tests while ignoring the smaller differences, inappropriate because the tabled values of t are based on the assumption that *chance*, not the selective eye of the experimenter, has operated freely. It should be quite obvious that one hundred independent t tests of a *true* null hypothesis will result in five that are significant at the 0.05 level, on the average.

A variety of tests have been devised which attempt to take into account the number of possible pairs of means in each experiment and which typically employ the overall within-group variance (i.e., based upon all of the groups) as an estimate of error. The best known of these "multiple comparison" tests of differences between all possible pairs of means are the Duncan test (Duncan, 1955) and the Tukey test (Ryan, 1959). Tukey's test is an attempt to use the error rate per experiment instead of per comparison as a basis for making significance statements, while Duncan's test compromises between these extremes and uses the error rate per *independent* comparison (per *df*) (cf. Ryan, 1959). The Tukey test is the more conservative, resulting in rejection of fewer null hypotheses. Possibly for this reason, it is also the less often used. On the other hand, Duncan's test treats "independent" comparisons within a single experiment as if they were independent in the sense that comparisons from completely different experiments are. It, thus, may be too lenient. A number of similar multiple comparison methods have been developed (most notably the Scheffé and Newman-Keuls tests, Winer, 1962), with varying degrees of conservativeness. It is likely that many of these methods will continue to be employed for some time to come, before some kind of convention is established on a completely rational basis.

The question is occasionally raised whether it is legitimate to undertake postmortem comparisons when the overall F test does not achieve statistical significance. On the assumption that the method employed is one which takes into account all of the possible comparisons between pairs of means (such as the Tukey test), there seems to be no persuasive argument against such a procedure. Since recent discussions of multiple comparison methods have convincingly established that selected t tests are inappropriate in any case, it would appear that the newer methods should be used *whether or not* the overall F was significant. On the other hand, there is nothing to prevent an experimenter from making a particular comparison which is implied a priori by his analysis of the problem, so long as his statistical test allows for *all* of the possible comparisons that could be made in relation to the same experimental hypothesis.

For example, even though the overall F test of the several groups' means is not significant, a comparison of a single control group with all of the other groups combined would surely not be precluded on any statistical basis. On the other hand, separate comparisons of the same control group with each of the experimental groups in turn could not be assumed to meet Duncan's criterion of independence, especially to the extent to which significance in any one of these separate comparisons would be taken as evidence in support of the same general experimental hypothesis. A test designed especially for this purpose, however, is available (Dunnett, 1955). Similarly, there is no reason that an experimenter cannot conduct an analysis of variance using only some of the various groups originally run but omitting others (such as analyzing the data of only the groups receiving different degrees of some treatment but omitting the group receiving no treatment whatever).

On the other hand, segregation of the experimental data into analyzed and unanalyzed parts (or separately analyzed parts) must always take into account the increased opportunity of rejecting a true null hypothesis that may result from multiple testing. It remains the experimenter's responsibility at all times to report as accurately as he can the probability that chance alone may have been responsible for observed differences and to employ methods that are appropriately suited to his actual experimental procedures.

Orthogonal Contrasts

A qualitative independent variable may sometimes lend itself to qualitative subclassification, so that half of the data may be compared in a meaningful way with the other half. For example, in a maze running experiment testing four different food substances as incentives, the overall F may prove to be short of significance. If two of the substances, say, consisted largely of carbohydrate materials, there would again be no serious argument against a comparison of the combined carbohydrate conditions with the combined noncarbohydrate conditions. Actually, a set of four groups yields three independent contrasts of half of the data with the other half, each of which, as will be shown in connection with the discussion of *ordered* independent variables, may be tested separately. Since the likelihood of this situation arising in an experiment involving qualitative independent variables is actually very slight, we will consider it in connection with quantitative factors.

Suppose, for example, that instead of four food substances that differ qualitatively, four different concentrations of the same substance define the independent variable (one of these could be zero concentration, of course). Table 8.2 shows the way in which three independent contrasts of *two groups versus two groups* may be generated in this type of experiment. In Table 8.2 the columns represent the four different concentra-

TABLE 8.2

Orthogonal set of three comparisons in an experiment involving an independent variable with four ordered levels.

	Amount of concentration			
Contrast	a_1	a_2	a_3	a_4
I	+	+	−	−
II	+	−	−	+
III	+	−	+	−

tions and the rows indicate the three separate tests. The plus and minus signs (actually plus and minus 1) in the table indicate which pairs of groups combine for each test, as described below.

The three *df* associated with the differences among the four groups' means are partitioned in Table 8.2 into three contrasts, each involving a single *df*. That these contrasts are, in fact, independent, or orthogonal, to one another is established by noting that the sum of the cross-products of the orthogonal coefficients (the plus and minus signs) equals zero for each possible pair of two rows. Another way of putting this is to note that the first contrast, in which a_1 and a_2 are pitted against a_3 and a_4, is *balanced* with respect to the other two contrasts (i.e., contains a plus and a minus from each in both its positive and negative sides). This is true of each of the three contrasts with respect to the others.

This independence of the three contrasts can be seen also in the fact that the first one involves a comparison of the two groups that received smaller concentrations, on the one hand, with the two that received larger concentrations, on the other. The second contrast compares the extreme pair with the middle pair. The third contrast would appear to reflect some kind of oscillation effect, since it involves a comparison of the first and third amounts of concentration with the second and fourth amounts. Each of these possible effects could be present significantly or absent without quantitatively influencing the presence or absence or the size of the others, another meaning of the term "independence." Thus, the significance of any one of these orthogonal effects would not be relevant to the same experimental hypothesis involving any of the others.

Computation of variance estimates for these orthogonal contrasts may be achieved by means of the following formula:

$$\frac{(\Sigma kT)^2}{n\Sigma k^2}$$

in which T refers to the sum of the dependent measures within each group, k refers to its appropriate coefficient, as shown in the table ($+1$ or -1), and n is the number of measures (subjects) in each group. The sum of the three separate variance estimates (each is also a sum of squares, since each involves only a single *df*) equals the sum of squares among the four conditions, with 3 *df*. The estimate of error variance based on the within-groups variance is employed in the denominator of each of the three F tests, as well as in the overall F test. Table 8.3 shows the breakdown of *df* for this analysis of variance.

Trend Analysis

When the independent variable is not only quantitative, but is administered in *equal* quantitative steps (or equal logarithmic steps) the *trend* of the mean dependent variable measures may be analyzed into single *df* components in a potentially more effective way, so that specific quantitative hypotheses may be tested.

TABLE 8.3

Breakdown of *df* for analysis of variance employing orthogonal contrasts of separate *df*.

Source	*df*
A	$[a - 1]$ *
I $[a_1 + a_2$ vs. $a_3 + a_4]$	1
II $[a_1 + a_4$ vs. $a_2 + a_3]$	1
III $[a_1 + a_3$ vs. $a_2 + a_4]$	1
Error	$a(n - 1)$

* Brackets are placed around the *df* for *A* because they are repeated separately in the three orthogonal contrasts.

In the case of a four-group experiment, concentration of sucaryl might be varied in four equal quantitative steps and a measure obtained of preference behavior of rats (in which each concentration is presented in a container alongside of a container of water without any sucaryl). If the experimenter were interested especially in the hypothesis that preference increases as sucaryl concentration increases, and if his highest concentration exceeded an amount that was optimal for the animals in his study, data such as are depicted in Figure 8.1 might render his interpretation of the overall *F* somewhat difficult. The overall *F* is, after all, dependent upon any and all manner of variation among the four means. However, the trend of the preference means across amount of concentration may be analyzed separately with respect to its linear, quadratic, and cubic components (the specific meaning of each of these is considered below).

The method is similar to that employed in making orthogonal contrasts, except that the coefficients used in conjunction with each sum of measures are obtained from a table of coefficients of orthogonal polynomials (e.g., Fisher and Yates, 1953) for separately and sequentially fitting the linear, quadratic, and cubic components of the regression of the dependent variable on the independent variable. An abbreviated version of this table may be found in the Appendix.

It is possible, using this method, to evaluate as many higher order components as there are *df* among the entire set of means. In the case of the present example with four groups, the coefficients for each separate component are shown in Table 8.4. Also given in the table are the sums of the squared coefficients, since these are needed for computing the variance estimates. Substituting the appropriate coefficients in place of the *k* values in the formula on page 162, the variance estimates are com-

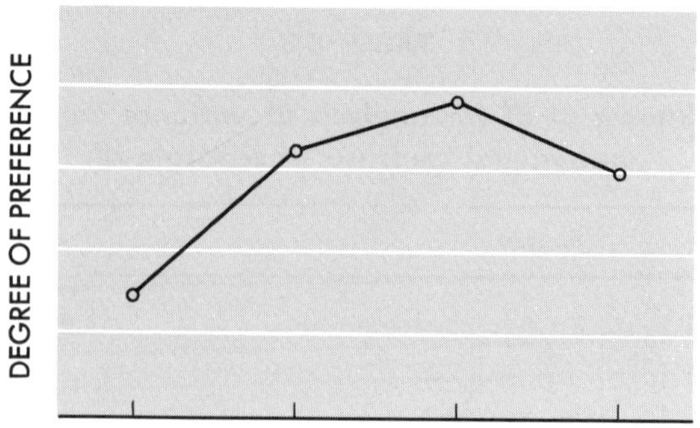

Figure 8.1 Hypothetical data showing mean preference as a function of concentration of sucaryl, varied in equal quantitative steps.

puted in the same way as was done for the conceptually more simple orthogonal contrasts. Examination of the sets of coefficients in Table 8.4 will indicate that each component (i.e., linear, quadratic, and cubic) is orthogonal to the others by virtue of the sum of the cross-products of the coefficients being equal to zero (e.g., for the linear vs. quadratic pair, $(-3)(1) + (-1)(-1) + (1)(-1) + (3)(1) = 0$).

TABLE 8.4

Coefficients of orthogonal polynomials for evaluating the linear, quadratic, and cubic components of the trend of preferences as a function of concentration of sucaryl, as depicted in Figure 8.1.*

Component of trend	Concentration of Sucaryl 1	2	3	4	Σk^2
Linear	−3	−1	1	3	20
Quadratic	1	−1	−1	1	4
Cubic	−1	3	−3	1	20

* The coefficients in each row are appropriate for determining the best-fitting linear, quadratic, or cubic equation.

Each separate component is again tested individually using the within-groups variance to estimate error. In the present example, the test of the linear component of the trend of the preference means would indicate whether the upward drift of the four means was significant, the test of the quadratic component of the trend would indicate whether the inverted U-shape in the data was significant, and the test of the cubic component would indicate whether significant oscillation existed (the last being the least psychologically meaningful test in this situation).

In a manner of speaking, these tests of trend successively answer the questions of whether the best fitting linear equation accounts for a significant portion of the variation among the means, whether the best fitting quadratic equation accounts for a significant portion, and, finally, whether the best fitting cubic equation accounts for a significant portion. In the present case, as in the case of orthogonal contrasts, the sum of the three variance estimates, each with 1 *df*, equals the sum of squares among the four means, with 3 *df*.

FACTORIAL EXPERIMENTS

The term "factorial design" is technically reserved for experiments involving more than one independent variable, in which the several independent variables (or *factors*) are orthogonal to one another. In this case, nothing more is meant by "orthogonal" than that each of the levels of each factor is administered in combination with each of the levels of the others. Each factor is a genuine independent variable in its own right, rather than, say, some factors being control variables that are run solely to reduce estimates of error or otherwise to improve the chances of obtaining unambiguous information about the effects of other factors. Experiments that are factorial in form but do not meet the technical criterion of involving at least two *genuine* independent variables with effects of interest may be referred to as *quasi-factorial* experiments. In many, if not most, actual applications of *quasi-factorial* designs, the actual computational procedures and analytical logic are quite the same as in genuine factorial experiments. Except as noted below, no further differentiation will be made between them.

Advantages of Factorial Experiments

Because the independent variables in a factorial experiment are orthogonal to one another, this type of experiment permits more efficient use of what are ordinarily limited supplies of subjects. A factorial experiment involving, say, three independent variables is somewhat the same as three separate experiments, each studying a different one of these variables. In the case of three separate experiments, however, three different sets of subjects would be employed, while the factorial experiment requires a single sample of subjects which may be only slightly larger than that of any one of the separate experiments. As will be shown later, the evaluation of each independent variable's effect uses *all* of the subjects in the factorial experiment, but the same subjects are involved each time.

A more important advantage of factorial experiments, even with respect to conclusions regarding the effect of just one of the independent variables, is that they may considerably broaden the range of generalizability of these conclusions. Instead of having held "everything constant," as in the case of single-factor experiments, additional relevant variables have been introduced, each at two or more different levels. Thus, one is not confined to some constant level of each of these relevant variables in making statements regarding the effect of the single variable in question. Instead, the experimenter may be in a position to conclude that the effect in question occurs similarly across several levels of the other variables or, perhaps, that the effect is different at different levels of one or another of these variables.

Simple and Main Effects

In a factorial experiment, the effect of one of the independent variables at a single level of another of the variables is referred to as a simple effect (e.g., the simple effect of A at B_1). The overall effect of an independent variable, averaged across all of the levels of the other independent variables, is referred to as its *main* effect. In a single-factor experiment (where the other relevant variables are held constant while one variable is varied systematically), it is as if the simple effect of the independent variable, at the levels of the other variables that were chosen to be held constant, were being studied. The manner in which the simple effects of a variable may differ from level to level of other variables is what is meant by the term "interaction." If the simple effects of a variable, A, differ significantly from level to level of another variable, say, B, this is described by stating that the interaction between A and B was significant. When the simple effects of A differ significantly across the levels of B (i.e., when there is a significant $A \times B$ interaction), it is also true that the simple effects of B differ significantly from level to level of A. The interaction of A and B implies *both* of the foregoing; it implies both that the effect of A depends upon B and that the effect of B depends upon A.

2 x 2 Design

To further clarify the nature of these relationships, let us consider an experiment involving two orthogonal factors, A and B, each with two levels (this is called a 2 x 2 design or a 2^2 design). For simplicity, we will assume that A represents the presence or absence of some treatment and that B represents the presence or absence of some other treatment (although it is also possible that A or B could represent different degrees of treatments or different types of treatments). Those treatment-combinations for which A is present will receive the label a, and those for which B is present will receive the label b. Figure 8.2, thus, shows the plan of this 2 x 2 factorial experiment and indicates the manner in which these labels may be employed to identify the four combinations of levels of A and B. In Figure 8.2, the cell involving both A and B absent (i.e., the

		B	
		b	—
A	a	ab	a
	—	b	(1)

Figure 8.2 Plan of 2 × 2 factorial experiment showing method of identifying the four separate treatment-combinations.

control condition) is arbitrarily labeled (1), for reasons which will be made clear in the section on the 2^n family of factorial experiments. In addition, we will assume that subjects are randomly assigned to the different treatment-combinations, with n subjects in each combination.

The main effect of A is contained in the difference between the marginal row means (i.e., mean of cells ab and a combined versus cells b and (1) combined); also, the main effect of B is given by the marginal column means (i.e., cells ab and b combined versus cells a and (1) combined). The simple effect of A with B present involves the difference between the ab and b cell means, while the simple effect of A with B absent involves the difference between the a and (1) cell means. The difference between these two *differences* defines the A x B interaction,

$$(\bar{X}_{ab} - \bar{X}_b) - (\bar{X}_a - \bar{X}_{(1)}).$$

Expressing the same interaction in terms of the difference between the two simple effects of B at the two levels of A results in a superficially different statement,

$$(\bar{X}_{ab} - \bar{X}_a) - (\bar{X}_b - \bar{X}_{(1)}).$$

Each of these descriptions of the A x B interaction simplifies algebraically to

$$(\bar{X}_{ab} + \bar{X}_{(1)}) - (\bar{X}_a + \bar{X}_b).$$

indicating that the A x B interaction consists of a contrast of the two pairs of treatment-combinations which lie opposite from each other on the diagonals of the experimental plan.

It is easy to see that the main effect of A (which involves a contrast of half of the data with the other half) is *balanced* with respect to the B factor, since each level of A is run half with B present and half with B absent. Likewise, the main effect of B (also involving a contrast of half of the data with the other half) is balanced with respect to A. The $A \times B$ interaction is similarly balanced with respect to both A and B, since each diagonal pair of conditions is contributed to, half by A present and half by A absent, as well as half by B present and half by B absent. In other words, the effects which involve contrasts of half of the data with the other half (A, B, and $A \times B$) are all orthogonal to one another.

This relationship may be illuminated additionally by constructing a table similar to Table 8.2, which was used to illustrate orthogonal contrasts in a single-factor experiment involving four different treatments. Thus, Table 8.5 shows the orthogonal arrangement of the three main

TABLE 8.5

Orthogonal set of main contrasts in a 2 x 2 factorial experiment.

	Condition			
Contrast	***ab***	***a***	***b***	**(1)**
A	+	+	−	−
B	+	−	+	−
$A \times B$	+	−	−	+

contrasts in the 2 x 2 factorial design. Computation of sums of squares for the three orthogonal effects in the 2 x 2 factorial experiment is achieved by means of the same formula employed in the case of orthogonal contrasts in a single-factor experiment, given on page 162. Again, the appropriate error estimate for evaluating the significance of each of these effects is the within-group variance.

2^n Factorial Experiments

The 2 x 2 factorial experiment we have been considering is actually the simplest example of an entire family of experiments involving independent variables with only two levels or categories. When each of the variables in the experiment has only two levels, the total number of treatment-combinations, or conditions, equals 2^n, which gives this type of experiment its name. In experiments of the 2^n family, each of the main effects and interactions consists of a contrast of half of the total data with the other half. The entire set of such contrasts can be shown to be orthogonal in the same way as was illustrated in Table 8.5.

For example, we will consider a three-factor experiment, or a 2^3 experiment. As has already been indicated, the orthogonal combination of each of the levels of each of the three factors generates 2 x 2 x 2, or eight, different conditions. We will assume that n subjects have been assigned randomly to each of these conditions and, thus, that there are $8n$ observations in all. The segregation of these eight conditions into two subgroups of four each, to identify the contrast which defines each of the main effects and interactions, is achieved by performing the multiplication indicated by the polynomial expression

$$(a \pm 1)\ (b \pm 1)\ (c \pm 1),$$

with a minus sign following the letter associated with the effect in question and plus signs following the other letters (in the case of interactions, minus signs are used following each of the letters participating in the interaction and plus signs follow the excluded letters). Thus, for the effect of A, the polynomial becomes

$$(a - 1)\ (b + 1)\ (c + 1),$$

which reduces to

$$abc + ab + ac + a - b - c - bc - 1.$$

Thus, the effect of A involves a contrast of the four conditions with A present, with the four conditions with A absent. In a similar way, the B effect involves a contrast defined by

$$abc + ab + bc + b - a - c - ac - 1,$$

and the C effect is defined by

$$abc + ac + bc + c - a - b - ab - 1.$$

The A x B interaction, as was shown above, requires that minus signs follow two letters in the polynomial

$$(a - 1)\ (b - 1)\ (c + 1),$$

which produces

$$abc + ab + c + 1 - a - b - ac - bc.$$

Similarly, the B x C interaction is defined by

$$abc + bc + a + 1 - b - c - ab - ac,$$

the A x C interaction is defined by

$$abc + ac + b + 1 - a - c - ab - bc,$$

and the $A \times B \times C$ interaction, with minus signs following all three letters in the polynomial, is defined by

$$abc + a + b + c - ab - ac - bc - 1.$$

Table 8.6 illustrates that the entire set of seven contrasts are orthogonal with each other.[3]

TABLE 8.6

Orthogonal set of contrasts in the 2 x 2 x 2 factorial experiment.

	Condition							
Contrast	***abc***	***ab***	***ac***	***a***	**bc**	**b**	**c**	**(1)**
A	+	+	+	+	−	−	−	−
B	+	+	−	−	+	+	−	−
C	+	−	+	−	+	−	+	−
$A \times B$	+	+	−	−	−	−	+	+
$A \times C$	+	−	+	−	−	+	−	+
$B \times C$	+	−	−	+	+	−	−	+
$A \times B \times C$	+	−	−	+	−	+	+	−

INTERPRETATION OF INTERACTION

Beyond the essentially symbolic definition of interaction outlined so far, its meaning and its value to experimenters may be clarified in a number of additional ways. One of the most effective approaches to understanding what interaction involves may be accomplished by means of a few graphical examples. Figure 8.3, accordingly, depicts the two different ways in which the same two-factor interaction can be presented graphically.[4]

Each of the graphs shown in Figure 8.3 is based upon the same four means. In the left-hand graph, anxiety constitutes the independent variable, the dependent variable is the mean number of correct responses, and the factor (or *parameter*) defining the difference between the solid and dashed lines is intelligence. In the right-hand graph, intelligence is the independent variable and anxiety is the parameter in the figure. It should be quite obvious, then, that the meaning of the anxiety x intelli-

[3] Sums of squares are again computed by the formula on page 162.
[4] Data shown in these figures are from Denny (1966).

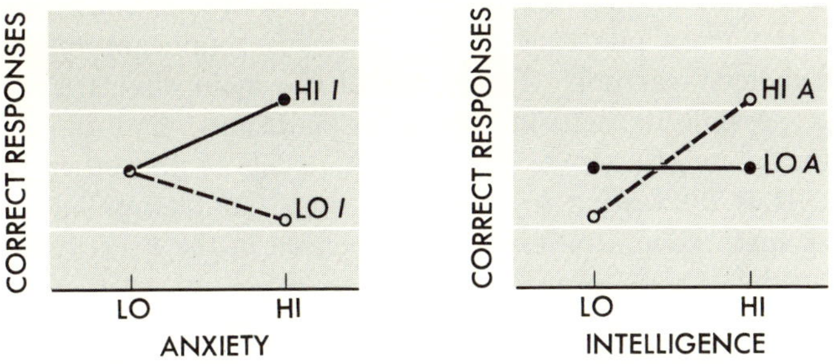

Figure 8.3 Two different graphical representations of the same two-factor interaction.

gence interaction in this situation may be verbally expressed also in at least two different ways. Referring to the left-hand presentation, it may be said that the effect of anxiety on concept performance is opposite in direction in the low intelligence subjects from what it is in the high intelligence subjects, since an increase in anxiety was associated with a *decrement* in performance in the low intelligence subjects but was associated with an *increment* in performance in the high intelligence subjects. The decision to present anxiety as the independent variable in the figure on the left has the effect of *focusing* interpretative considerations upon that variable. In relation to the right-hand presentation, one could state that the effect of intelligence on concept performance is confined entirely to the high anxiety subjects and is totally absent in the low anxiety subjects. Here the interpretative focus is on the intelligence independent variable.

How To Focus Interpretation

The question immediately arises whether there is any convention for choosing to employ *one* rather than the other of two equally possible and plausible representations of the results. The author of the study in question concluded from his data that anxiety was an important factor in concept performance only for the high intelligence subjects, indicating in this way that his preference was for focusing on anxiety and treating intelligence as a parameter. Since neither anxiety nor intelligence was a manipulated independent variable (i.e., the study was *quasi-experimental* in form), there would appear to be little to choose between the two possible approaches.

One consideration which sometimes may steer the experimenter in one interpretative direction as opposed to another is whether either of the main effects (e.g., in the current example, anxiety or intelligence) was itself separately significant. In the study under consideration, for example, the main effect of intelligence was significant but the main effect of anxiety was not. If only *one* of the main effects is significant, it may be possible to focus interpretation on it and relate the interaction to changes in its effect over the different levels of the other factor (or factors). It is interesting to note that, even though intelligence had a significant main effect, the experimenter chose in the present instance to emphasize the anxiety variable.

Obviously, if one of the variables entering into an interaction is experimentally manipulated while the other is varied by selection of subjects, the choice for focused interpretation of the interaction presents no problem whatever; the manipulated variable should be emphasized. In the quasi-experiment in question, *both* variables were selected rather than manipulated. Thus, the experiment's choice of focus was entirely arbitrary.

A Priori Bases of Focusing

When both of the variables in an interaction were experimentally manipulated, there may still be a priori logical reasons for preferring one direction of focus over the other. Many times the experimenter's theoretical line of reasoning almost automatically moves in the direction of one of the independent variables rather than the other. For example, suppose an experiment were conducted on the roles played by pre-experimental instructions and the presence or absence of a ready signal in acquisition of classical eyelid conditioning. Suppose further that the instructions variable were implemented by telling one group of subjects to attempt to inhibit making anticipatory conditioned responses while the other group was told to facilitate making anticipatory responses. Half of each of these instructions groups are run with a ready signal before the beginning of each trial and half are run without a ready signal. Suppose, finally, that a significant interaction between instructions and ready signal were found (along with a significant main effect of instructions *and* a significant main effect of ready signal). Figure 8.4 presents a graphical representation of these results and illustrates the direction that would probably be taken in focusing the interpretation of the significant interaction. Notice that the independent variable in the figure is the instructions factor, while the parameter is the ready signal.

The experimenter could interpret this interaction in somewhat the following way, "The effect of the instructions variable was enhanced by the presence of a ready signal." Since, even without a ready signal, the facilitative and inhibitory instructions appeared to generate differences in level of conditioned responding, presumably by lowering it in the

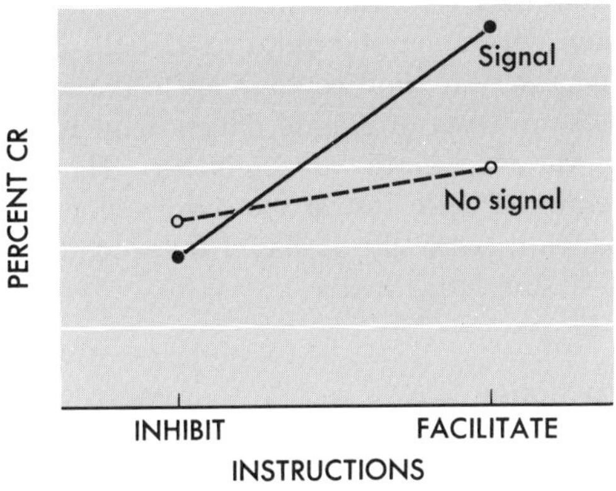

Figure 8.4 Hypothetical data showing focused interpretation of significant interaction between instructions and ready signal in eyelid conditioning.

group that received inhibitory instructions and raising it in the group that received facilitative instructions (an uninstructed control group would be useful in this connection), the role of the ready signal would seem to be best understood as that of an amplifier of the instructions effect. It would be somewhat clumsy to interpret the interaction by focusing on the effect of the ready signal, since the *direction* of the conditioning effect is not determined by it but by the instructions variable. Even so, it would still be possible to conclude, "The effect of the ready signal depended upon whether the subject had received facilitative or inhibitory instructions."

From the present point of view this verbalization, while possible, does not seem to speak as well to the most probable underlying processes as does the statement presented in the previous paragraph, focusing on the instructions variable. Also, while the main effect of the ready signal *was* significant (higher level of conditioned responding with a ready signal present than without one), this was due entirely to the very high level of responding in the group that had facilitative instructions along with the signal. As must be clear from this discussion, the experimenter must decide which interpretative focus to adopt, and his discharge of this responsibility must conform to the explanatory directions already developed in his initial approach to the problem, except, of course, when the results are totally unexpected.

Interaction in Fixed Constants Experiments

When the interaction between two variables is significant in a random groups factorial design, and when the factors entering into the interaction are both of the fixed type (cf. Chapter 7, p. 144), interpretation of either of their main effects may be quite restricted. In situations of this type it becomes necessary to limit conclusions to the *simple* effects of the variables in question. Tests of simple effects may be done, employing the overall error variance in the denominator. The variance estimate used in the numerator of an F test of the simple effect of a factor at one level of another factor, naturally, is based only on the data within the appropriate level of the latter factor.

Relational Interactions

In certain factorial experiments, a significant interaction may not lend itself to focused interpretation with respect to *either* independent variable. These occasions arise when the separate factors have only trivial implications themselves, but in combination define a complex, or relational theoretical concept. To exemplify this type of interaction, let us assume that an experiment is conducted in which four different groups of randomly assigned subjects receive training involving four different cue stimuli, the stimuli being small patches of light which differ only in color (i.e., which differ in dominant wave length). For this purpose we could employ, say, blue, green, yellow, and red appearing lights. At the end of training the four groups of subjects are subdivided for testing, each original group being broken randomly into four subgroups, each of which is tested with one of the four colored lights. The plan of this factorial design is shown in Figure 8.5.

	Test Stimulus			
Training Stimulus	Blue	Green	Yellow	Red
Blue				
Green				
Yellow				
Red				

Figure 8.5 Plan of factorial experiment on stimulus generalization.

It is quite possible that the experimenter would have little interest in either the main effect of stimuli used during training or of stimuli used during testing. The interaction, however, may be of considerable significance to him because of the possibility that it may reflect a gradient of stimulus *generalization*. The subjects who are tested on stimuli that are identical, or highly similar, to the stimulus used during training would be expected to show a greater tendency to make the learned response than would subjects who are tested on stimuli more different from the originally trained stimulus. A significant interaction would not, in this case, result in an interpretation focused on one or the other of the two independent variables (e.g., "The effect of differences among stimuli used during training depends upon which stimulus was used during testing," or "The effect of differences among test stimuli depends upon which stimulus was used during training"), even though either of these statements would be technically accurate.

The use of factorial interactions to assess the presence or the extent of relational theoretical processes, while not different in procedural or quantitative detail from their less subtle applications, would appear to open a door into far more complex areas of experimentation than have heretofore been studied successfully. Beyond the more obvious topics, such as *generalization, transposition,* and *transfer,* which may be amenable to such an approach, relational concepts in interpersonal situations (e.g., clinical *transference*), perceptual situations (e.g., *figure-ground, perceptual disparity*), and others may lend themselves to similar approaches. An excellent example of the potential *unique* effect of particular combinations of levels of two independent variables is given by Box and Youle (1955). It would seem that only to the extent that explanatory theory remains undeveloped is a particular research area beyond the scope of this device.

Successive and Simultaneous Interactions

The interaction discussed above in connection with relational concepts (depicted in the plan of Figure 8.5) differs from those previously considered in yet another way. In the generalization example, the *times* of administration of the two different independent variables were different, since the first variable involved differences among stimuli used during training while the second involved differences among stimuli following training. This type of interaction may be referred to as *successive,* differentiating it from *simultaneous* interactions in which the participating variables are administered together. Although both successive and simultaneous interactions may reflect complex relational concepts, only the successive type of interaction requires a change in conditions for the subjects in the study. Beyond the kind of interaction exemplified in the generalization experiment (defined in relation to differences between training and testing cues), an additional effect due solely to change itself (e.g., surprise) may be manifested in the first or second test trial. Thus, it may happen that an experimenter will omit test trial scores immediately following a change in conditions, using only the subsequent test data to evaluate his hypotheses.

In Chapters 9 and 10 a variety of experiments will be considered in which double-classification resembling the factorial designs is employed. These experiments are basically of the type previously referred to as quasi-factorial, in that the added variable (the blocking variable) is not being studied in its own right, but is used solely for the purpose of increasing sensitivity in the tests of the other variables. Since analysis of variance is fundamentally nothing more than a highly efficient computational scheme for partitioning variability into its additive components, the mechanics of partitioning in these experiments are quite the same as those employed in factorial experiments. The difference between such designs and the factorial design is perhaps best illustrated by the application of the principle of blocking to the factorial design itself, also described in Chapters 9 and 10. In such cases, it will be seen that the blocking variable serves an entirely different purpose from that served by genuine independent variables in factorial experiments.

Further, in Chapter 10, a number of experiments are considered in which the individual subject serves as the unit of blocking, receiving all of the levels of each of the independent variables. While it is common in such experiments to speak of, say, a treatments x subjects interaction, the similarity of this type of interaction to those considered in this chapter ends with their algebraic and computational characteristics. These points will be made somewhat clearer in the next two chapters.

SUMMARY

The *random groups* design employs different groups of subjects, assigned randomly to the several levels of the independent variable. The basic logic for testing the null hypothesis is that the *pooled* variability of subjects who are treated similarly provides an estimate of error for evaluating the observed variability among the groups' means. A single *replication* of an experiment of this type would contain one subject in each condition. The differences among these subjects define the *phenomenon* under study, thus the within-groups variance estimate (i.e., replication of the phenomenon) provides the appropriate error term.

The pooling of the separate within-groups variances requires the assumption that they be homogeneous. However, if the number of subjects is approximately equal from group to group, violation of the assumption of *homogeneity of variance* will produce little change in obtained probabilities. Extreme instances of heterogeneity may sometimes be dealt with effectively by *transformation* of the dependent variable.

Multiple comparisons of pairs of groups following an overall analysis requires special tests which account for the number of such comparisons being made. *Orthogonal contrasts* of single-*df* effects may also sometimes be desirable. When the independent variable involves equal steps on a quantitative scale, analysis of the separate components of the overall *trend* may also be highly useful for testing special quantitative hypotheses.

The same general logic applies to random groups designs involving *factorial* combinations of levels of two or more variables. Experiments of this type have the advantage of economy, they tend to broaden the range of applicability of the results regarding each of the separate factors, and they permit the study of *interactions* of variables. Interpretation of interaction may depend upon the structure of the experimental design, in that the *focus* may be on actually manipulated factors instead of selected variables, or on the actual experimental data, in that the focus may be on factors whose *main effects* were also significant. There may also be a priori theoretical reasons for focusing on one or another of the component factors. In the random groups factorial design, a significant interaction frequently makes it necessary to limit conclusions to the *simple effects* of the variables involved, since the interaction means that the simple effects of one of the factors differ from one level of the other factor to the next. Sometimes a significant interaction may reveal the operation of a *relational* theoretical process which could not be studied in single-factor studies.

9

The Blocking Principle

HETEROGENEITY OF THE EXPERIMENTAL MATERIAL

The conclusions hopefully to be drawn from the results of an experiment are usually intended to have applicability for a large group of subjects of a particular type or class. Yet, each individual observation that is made must be made on a single member of the group. This is true even when more than one observation may be made on each member. Ideally, we would wish to be able to observe the effect of an independent variable (across all of its levels or categories) within a hypothetical individual who is a representative member of the entire class.

As should be obvious, however, this ideal exists only conceptually. In the real world of experimentation, it is usual for the individuals who serve as subjects in our experiments (the experimental material) to differ in the behavior to be observed almost as much beforehand as the different experimental treatments they receive may make them differ afterward. Under these conditions, only extremely large experiments (with many subjects in each condition) would achieve sufficient sensitivity to detect even important experimental effects.

The basic method for dealing experimentally with the general problem of heterogeneity of the experimental material involves application of what is known as the *blocking principle*. Essentially, we must arrange to observe the basic experimental comparisons within blocks of experimental material (e.g., groups of subjects) that are more homogeneous regarding the behavior to be measured than is typical of the entire available sample. The more homogeneous, the better. Each block will be as large as is minimally necessary for all of the levels or categories of the independent variable to be administered within it.[1] In other words, each block constitutes a complete replication of the basic experiment.

Between blocks, the experimental material may differ greatly, since, as will be explained later, the larger the between-block variance, the more effectively has the blocking principle been applied. The blocking principle was originally developed in the field of experimental agriculture (Fisher, 1953) to deal with the problem of soil heterogeneity, and because its most elementary application is well-exemplified in that area, we will first consider the problem in an agricultural context.

RANDOMIZED BLOCKS DESIGN

Suppose that we are planning to conduct an agricultural experiment on the effect of the concentration of a fertilizer on the growth rate of a certain crop. A large rectangular field has been allocated for the purposes of this research. This field is subdivided into small plots, each sufficiently large to contain only a single growing plant while allowing adequate space between adjoining plots to ensure isolation of each from all the others.

The simplest procedure which might be used in assigning the different concentrations of fertilizer would be, of course, to apply the different concentrations to different large segments of the field, each segment containing several plots. But, this procedure would be rejected as unsatisfactory on the obvious grounds that the different segments may differ greatly in soil fertility and any observed differences in growth rate of the plants that received the different concentrations of fertilizer could just as easily be due to soil differences as to the independent variable. Nor is this problem solved by employing randomization to determine which concentration of fertilizer is applied to which large segment, since chance factors will still (even more explicitly) determine which large segment receives which amount of concentration of fertilizer.

Complete Randomization

A better procedure would be to use randomization for assigning different concentrations *to each separate plot* in the field. This would ensure that the error variance due to soil heterogeneity (i.e., differences in fertility of different plots of soil) would be distributed according to chance *both between and within* the different levels of the independent variable. This, it should be apparent, is the basis of the simple random

[1] Situations in which the size of the block is smaller than the size of a replication, exemplifying confounding, will be discussed in detail in Chapter 11.

groups design discussed on pp. 173–74. Under the null hypothesis, separate estimates of the same population variance (the variance of individuals or plots treated similarly) can be obtained from the means of different conditions and from the individual observations within conditions, so that a test of significance is available.

Unfortunately (because of the large amount of heterogeneity), the simple randomized design is what we would have to use if there were no way at all in which to know in advance which plots were likely to be more or less fertile. Suppose, for example, that the experiment involved six different concentrations of fertilizer and the available field were capable of being divided into 36 individual plots. Figure 9.1 illustrates

F-3	F-5	F-3	F-3	F-6	F-6
F-5	F-2	F-1	F-4	F-1	F-2
F-2	F-4	F-5	F-2	F-1	F-6
F-4	F-4	F-4	F-5	F-3	F-5
F-6	F-3	F-1	F-6	F-6	F-2
F-2	F-5	F-1	F-4	F-1	F-3

Figure 9.1 Plan of experiment on concentration of fertilizer, utilizing completely randomized assignment of six levels of concentration to the 36 separate plots.

one possible outcome of complete randomization of assignment of concentrations to plots. In the figure, the letter–number combinations refer to the different degrees of concentration of the fertilizer, F standing for "fertilizer."

Referring to the experimental plan shown in Figure 9.1, it is clear that a great deal of soil heterogeneity might inflate the error variance to a point that would render this experiment almost entirely futile. This is due to the fact that the appropriate error estimate consists of the pooled *within-condition* variance (the average variability of groups of plants that are treated alike), which is likely to be quite large in this case. Notice, for example, that the six plots containing plants that received the F-1 concentration of fertilizer are widely spaced in the field and, accordingly, are likely to differ substantially in growth rate in spite of having been treated similarly. This is, however, exactly what can be expected when a large amount of uncontrollable error variation is dealt with solely by randomization, rather than being reduced experimentally by blocking.

Reducing Heterogeneity

Fortunately, in the present case it is known that the major factors leading to soil heterogeneity are such as to make plots *near one another* more equally fertile than those separated by some considerable distance. Such things as the location of nearby streams, the distribution of natural chemicals in the soil, the angle of incidence of available sunlight, and the shading produced by trees and hills, etc., are all more likely to influence the fertility of adjacent plots more similarly than they would influence plots more widely separated. The simplest application of the blocking principle in this case would be to divide the field into six segments, or blocks, each of which contains six plots. Within each of these blocks, the six amounts of concentration of fertilizer are then assigned randomly, one to a plot (the reader will notice the difference between the method described here and the method mentioned on p. 179). The actual fashion in which the six blocks are formed may depend upon special features of the local environment and terrain. For example, if there is a stream running along one edge of the field, the blocks would be laid out as strips parallel to the stream, as is shown in the experimental plan in Figure 9.2. In other cases it might be preferable to use an arrangement of 2 x 3 rectangles, as is shown in Figure 9.3. Regardless of the manner in which the blocks are arranged, *within* each block the treatments are randomly assigned to the plots.

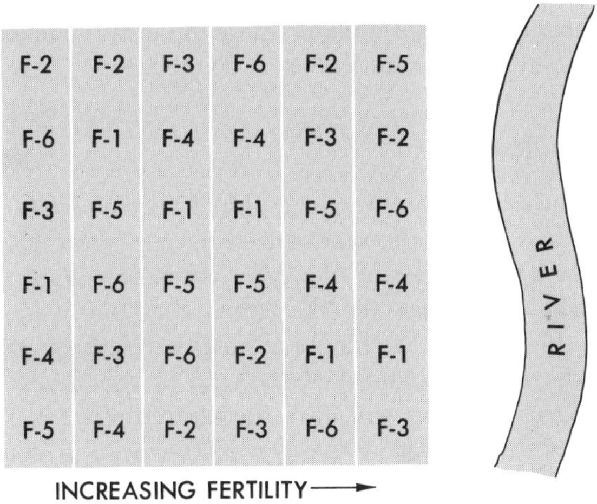

Figure 9.2 Plan of experiment on concentration of fertilizer, utilizing randomized blocks design to reduce error due to soil heterogeneity correlated with proximity of river.

F-1	F-5	F-4	F-1	F-5	F-6
F-6	F-3	F-6	F-3	F-4	F-1
F-2	F-4	F-5	F-2	F-3	F-2
F-5	F-6	F-5	F-2	F-4	F-2
F-3	F-1	F-1	F-3	F-6	F-3
F-4	F-2	F-6	F-4	F-1	F-5

Figure 9.3 Plan of experiment on concentration of fertilizer, utilizing randomized blocks design to reduce error due to soil heterogeneity arising from several sources.

Inspection of either Figure 9.2 or Figure 9.3 reveals that each block contains a complete replication of the experiment. The variability of growth rates *between* the six blocks is no longer a matter of concern in the test of significance, since the experiment is actually conducted *within* blocks and the appropriate source of error variance in this design arises from the opportunity for the effect of the independent variable, concentration of fertilizer, to differ from block to block. This variation, as should be clear by the previous statement, is the mean square for the conditions x blocks interaction. The more the blocks, as entire units, differ from one another, the more successfully has the design achieved relative reduction of error. Consideration of the breakdown of *df* for this analysis of variance and comparison of it with the appropriate breakdown for the analysis of variance of the simple randomized design will illustrate this point. These are shown in Table 9.1.

Increased Sensitivity

Table 9.1 shows that the randomized blocks design contains 5 *df* fewer for error than does the simple randomized design (more generally, $b - 1$ *df* fewer, where b = number of blocks). These are the *df* for the differences among the six blocks. To the extent that the average variability of the plots within each of the blocks is smaller, relatively, than the variability among the blocks, removal of the sum of squares for blocks from the sum of squares for error will have the desired effect of increasing the sensitivity of the experiment. Since the plots within a block are similar to one another in this agricultural example, this reduction is likely to be achieved.

It should be noted that once the decision has been made to conduct the experiment in randomized blocks, it becomes absolutely necessary to use

TABLE 9.1

Breakdown of *df* for simple randomized and randomized blocks designs compared.

Simple randomized design Source	*df*	***Randomized blocks design*** Source	*df*
Concentrations	5	Concentrations	5
Error (within conditions)	30	Blocks	5
		Error (concentrations x blocks)	25
Total	35	Total	35

the randomized blocks analysis of variance, i.e., to remove the variation due to blocks from the error estimate. Since the different levels of the independent variable were administered to experimental units (plots) that were more than ordinarily similar to one another, the calculated variance estimate for concentrations is likely to be smaller than it would have been had the simple randomized design method of assignment of concentrations to plots been used. This exemplifies a tangential feature of what is meant by the expression "the appropriate error term." Were the variance due to blocks not removed from the error estimate for analysis of variance of the data collected in a randomized blocks design, the probability of a Type 2 error would be seriously inflated. This problem will be discussed below in connection with matched pairs designs.

It will be recalled from Chapter 6 that the appropriateness of the error term in each different experimental situation may be determined on the basis of whether and by what means the *phenomenon* under study has been effectively replicated. In the randomized blocks design, as was noted above, the conditions x blocks interaction purports to meet this criterion of appropriateness.

The conditions x blocks interaction arises directly from the fact that the effect of the independent variable, concentration of fertilizer, may differ from block to block. Each block constitutes a *replication of the phenomenon* being studied, the *phenomenon* in this case being the way in which the concentrations variable influences growth rate (i.e., the *trend* of the six data-points in a particular block). In one block, for example, substantial increases in growth rate may have been associated with increases in concentration. In another block, the growth rates may have been relatively uninfluenced by the concentration of fertilizer. It may even occur that *lower* growth rates are associated with higher concentrations of fertilizer in a particular block.

Appropriateness of Error Estimate

To illustrate the fact that this variability (the conditions x blocks interaction) is exactly what will provide the appropriate error term in the randomized blocks design, data from two hypothetical experiments in randomized blocks are shown in Figure 9.4, each experiment involving three blocks. In one of these experiments, shown in the left-hand graph of Figure 9.4, considerable fluctuation in the influence of the independent variable among the three blocks is depicted. In the other experiment, shown in the right-hand graph of Figure 9.4, the trend of growth rates

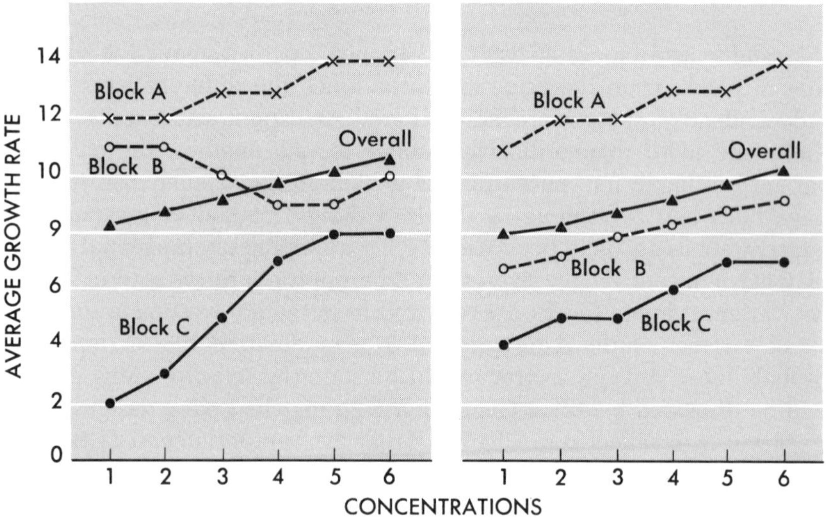

Figure 9.4 Two sets of hypothetical data illustrating the appropriateness of the conditions × blocks interaction as an error term in the randomized blocks design. Data from three blocks (*A*, *B*, and C) are shown in each graph. As is indicated, the average level of growth rate differs considerably between blocks in both experiments. In the left-hand graph, the conditions × blocks interaction is quite noticeable, but in the right-hand graph, it is negligible. The average overall trend across amounts of concentration of fertilizer is about the same in both graphs.

across concentrations is less variable from block to block. It should be noted that both sets of hypothetical data indicate substantial differences between blocks in the overall *level* of growth rate (i.e., blocking *was* used advantageously) and that the *average* trend of growth rate across concentrations (combining all blocks) was about the same in both experiments, as the two graphs show.

Even though the *average* effect of the concentration of fertilizer variable is approximately the same in the two graphs in Figure 9.4, the influence is far more consistent from block to block in the right-hand graph than in the one on the left. To the extent to which the different blocks *differ* in the way in which the independent variable has its influence, we should properly be less inclined to reject a null hypothesis that states that the independent variable has no influence. To the extent to which the effect of the independent variable is similar from block to block (and the trend is either upward or downward), we may be more confident of its reliability.

These considerations are the basis of the appropriateness of the conditions x blocks interaction as an error term to test the significance of the average effect of the independent variable. Assuming the blocks variable to be of the random type, in the terminology of Chapter 7 (p. 144), the conditions x blocks interaction contaminates the expected value of the variance estimate for the effect of conditions. Thus, in relation to the assumed underlying mathematical model, the interaction is the appropriate error term for evaluating the effect of conditions. Mathematical logic, accordingly, leads us to the same choice of error term in this situation, a conjunction which may be satisfying to those who prefer mathematical to intuitive argument.

PSYCHOLOGICAL EXPERIMENTS IN RANDOMIZED BLOCKS

The application of design principles developed in agriculture to psychological experimentation requires only a simple translation of concepts (Baxter, 1941). In experiments in psychology, as has been noted previously, the principle source of heterogeneity of the experimental material is what is often referred to as "individual differences." In relation to the design problem currently under consideration, this expression refers to differences *in the dependent behavior* which has been selected for experimental study.

If we are conducting an experiment in which the running speed of rats down a straight alley is to be measured, we must be concerned with the fact that rats differ in this measure whether or not we treat them differently. If we are planning to do an experiment on concept formation which requires that human subjects learn to make different responses to different classes of stimuli, we must be concerned with the fact that individuals differ in their ability to learn concepts even when they have not been treated differently.

Blocking by Grouping Subjects

In relation to our agricultural starting point, the psychological analogue of conducting the basic experiment within homogeneous blocks of soil involves the formation of *groups of subjects* who are likely to exhibit less variability in the behavior to be observed than is found among subjects in general. The experiment is then conducted separately within these homogeneous groups. Within each group, or block, the subjects are assigned randomly to the various levels or categories of the independent variable. While in agricultural research it is possible simply to assume that plots of soil that are spatially near one another are more likely to be equally fertile (i.e., to show similar growth rates when treated similarly) than are those that are more spatially remote from one another, very few such simple dimensions of *assured* similarity are available in the psychological analogy.

In psychological experimentation, existing knowledge regarding reliably distinguishable individual characteristics that are substantially *correlated* with the behavior to be measured must typically be made use of in forming effective blocks. The example of an agricultural randomized blocks design involving a known source of heterogeneity (a river), shown in Figure 9.2, is closer, in this way, to the psychological situation than any others used in agriculture. In the example shown in Figure 9.2, it was assumed that a correlation existed between soil fertility and proximity to the river, as was indicated by the arrow below the diagram of the field.

Natural Groups

It is sometimes possible in psychological experimentation to use natural groupings as the basis of a randomized blocks design, for example, in studies involving littermate controls or sibling groups. The only requirements to be met to ensure that the full benefit of blocking may be achieved in these cases are that the litters or sibling groups are each large enough to accommodate a complete replication, that the dependent variable measures differ considerably from litter to litter (or family to family), and that the different conditions be assigned to subjects at random within each block. The first of these requirements applies only when deliberate confounding is not intended (see Chapter 11 for a discussion of this topic). The second requirement simply reiterates the principle that the blocking variable (in this case litters or sibling groups) be correlated with the dependent variable.

If no correlation exists between the blocking variable and the observed behavior, an inefficient expenditure of wasted effort will have been made, and, in addition, there will be a slight reduction in the sensitivity of the experiment, since the number of *df* for error will be reduced by the removal of those *df* associated with blocks (thus, a larger F or t ratio would be needed for statistical significance). If there is no sensible reason for assuming that the dependent behavior is likely to vary less within litters than it does between litters (e.g., if the behavior is not thought to be highly genetically determined), it would be wasteful at best and harmful at worst to use litters as a basis for blocking. This is an admonishment to which we will return in discussing the matched pairs design in psychology.

FACTORIAL DESIGNS IN RANDOMIZED BLOCKS

The blocking principle is not limited to situations involving single-factor experiments, such as in the agricultural example considered above. Once it has been deemed necessary to conduct an experiment in randomized blocks and an appropriate blocking variable has been established (e.g., litters), it is possible to run all the different conditions of factorial experiments within blocks. For example, suppose that, instead of a single-factor experiment involving an independent variable with six levels, we wished to conduct a 2 x 3 factorial experiment involving six *combinations* of levels of the two separate factors. In the agricultural example discussed above, we might have been interested in three concentrations of fertilizer as one independent variable and two amounts of artificial irrigation as the other.

In using the randomized blocks design, it makes no difference whether the group of experimental conditions comprising a complete replication represents all the levels or categories of a single factor or all the combinations of levels of two or more factors. The only practical limitation in this regard is that the complete replication be small enough, relatively, so that it is feasible to conduct it within a single block (e.g., it might be quite impossible to obtain litters large enough to be used as blocks for, say, more than eight or ten combinations of conditions). Also, the larger the block, the less likely it is that homogeneity of the experimental material will be sufficiently great (in cases involving psychometrically defined blocking variables, as discussed below).

Figure 9.5 presents a plan for conducting a hypothetical factorial experiment on concentration of fertilizer and amount of irrigation in randomized blocks, and Table 9.2 shows the breakdown of *df* for the appropriate analysis of variance.

F-2 I-1	F-3 I-2	F-2 I-2	F-1 I-2	F-3 I-1	F-1 I-2
F-1 I-2	F-2 I-2	F-1 I-1	F-3 I-2	F-2 I-1	F-3 I-2
F-1 I-1	F-3 I-1	F-3 I-1	F-2 I-1	F-2 I-2	F-1 I-1
F-2 I-2	F-3 I-2	F-2 I-1	F-2 I-2	F-3 I-1	F-1 I-2
F-3 I-1	F-1 I-2	F-3 I-1	F-3 I-2	F-1 I-1	F-2 I-1
F-2 I-1	F-1 I-1	F-1 I-2	F-1 I-1	F-3 I-2	F-2 I-2

Figure 9.5 A factorial experiment involving three amounts of concentration of fertilizer and two amounts of irrigation, run in six randomized blocks. The three levels of concentration of fertilizer are identified by F-1, F-2, and F-3, while the two levels of irrigation are identified by I-1 and I-2.

Error Estimate

As has been noted previously, the error term for analyzing an experiment run in randomized blocks consists of the conditions x blocks interaction, although, in this case, the conditions x blocks interaction, itself, is a composite of concentrations x blocks (10 *df*), irrigations x blocks (5 *df*), and concentrations x irrigations x blocks (10 *df*). In using a pooled error term such as this, as was noted in Chapter 8, it is implicitly assumed that its components are homogeneous, i.e., that they are all estimates of the same error variance. Although in agricultural experiments in randomized blocks it is not common for this assumption of homogeneity to be explicitly tested, inspection of the separate components of pooled error terms and, occasionally, testing of their homogeneity are frequently done in psychological experimentation, especially when the subject serves as a block. In Chapter 10, which is concerned with situations involving repeated observations on the same subject (and, in which the *subject* is the unit of blocking), this point will be treated more fully and methods for partitioning the components of error in these cases will be presented.

Assuming homogeneity of the components of error, the effects of concentrations, irrigations, and the concentrations x irrigations interaction are all tested by means of the single, pooled error term. If the assump-

TABLE 9.2

Breakdown of *df* for analysis of 2 x 3 factorial design run in randomized blocks.

Source	*df*
Concentrations	2
Irrigations	1
Concentrations x irrigations	2
Blocks	5
Error (pooled conditions x blocks)	25
Total	35

tion of homogeneity is not justified, each of the three separate effects could be tested with the component of error appropriate to it, concentrations being tested with concentrations x blocks, irrigations with irrigations x blocks, and concentrations x irrigations with concentrations x irrigations x blocks. By logic previously employed to demonstrate the general appropriateness of using the conditions x blocks interaction, these separate components of error similarly can be shown to be appropriate to test the separate effects in question.

DOUBLE BLOCKING—THE LATIN SQUARE DESIGN

The randomized blocks design is most useful, as has been suggested, when an explicitly identifiable blocking variable exists which is known to be correlated with the dependent variable to be measured. A good illustration of this was given in the plan shown in Figure 9.2, where a river running alongside the experimental field could be assumed to produce systematic variation in soil fertility as a function of distance from it. Figure 9.3 presented a variation of the randomized blocks design which was suggested for use when nothing more was known beyond the fact that proximity of plots to one another might provide a basis for increasing homogeneity.

A better plan for such situations, in which no explicit factor exists as a basis for blocking or in which *two* independent known factors are available for blocking, would be to arrange the assignment of treatments to plots (or experimental units) such that each treatment appears once and only once in each row and once and only once in each column of a square, known as a Latin square. Within each row and within each column, however, randomization is still the basis of assignment, as will be explained below. Figure 9.6 illustrates the way in which two orthogonal random-

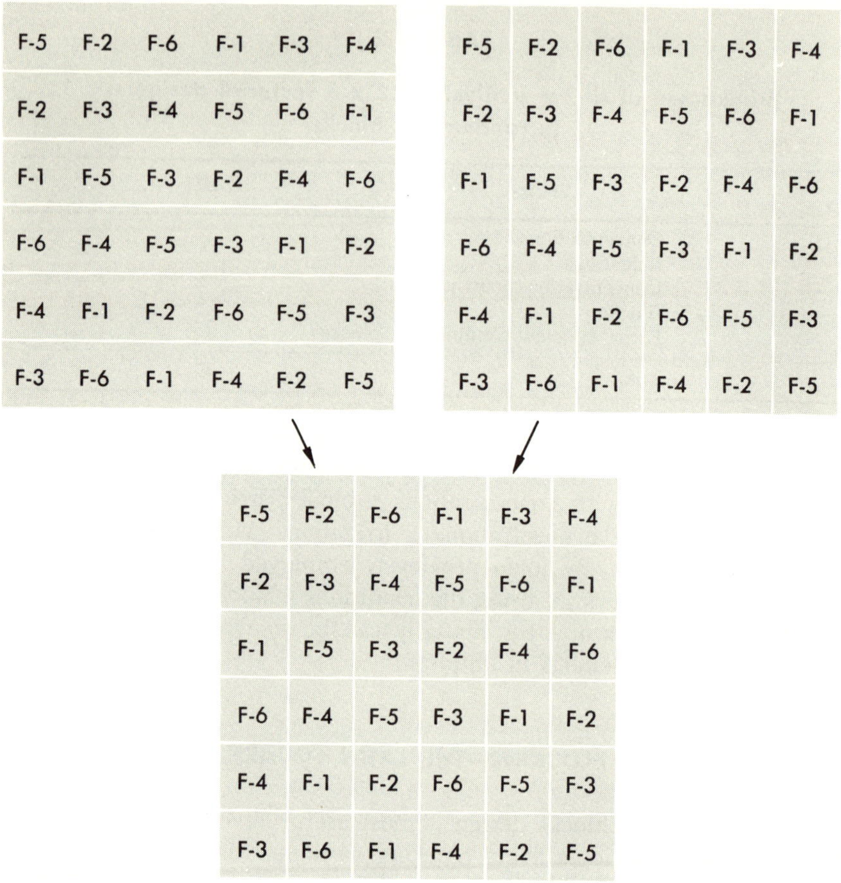

Figure 9.6 Combination of two orthogonal randomized blocks arrangements into a randomized Latin square design for the experiment on amount of concentration of fertilizer. The two orthogonal blocking arrangements at the top combine to form a double blocking at the bottom.

ized block arrangements combine to form a randomized 6 x 6 Latin square for conducting the experiment on amount of concentration of fertilizer.

Examination of the arrangement shown in the combined plan at the bottom of Figure 9.6 will confirm that each concentration of fertilizer appears only once in each row and only once in each column. Thus, if we imagine two independent influences (their independence here illustrated by the fact that they are placed at right angles to one another), such as a river running along one edge of the field and a large hill running along an adjacent edge, the rows represent randomized blocks (each containing a complete replication) that vary in distance from either the top or bottom edge, while the columns represent randomized blocks (each containing a complete replication) that vary in distance from either the left or right edge.

The name *Latin square* arises from the fact that Latin letters are typically used to identify the levels or categories of the independent variable on the "face" of the square. As will be shown below, more complex arrangements may involve the use of two or even three orthogonal "alphabets" to describe all of the combinations of treatments on the square.

Constructing a Randomized Latin Square

Perhaps the simplest and most efficient way to construct a randomized Latin square is to begin with a systematic square arrangement of Latin letters, such as is shown in Figure 9.7, and then randomly to rearrange

A	B	C	D	E	F
B	C	D	E	F	A
C	D	E	F	A	B
D	E	F	A	B	C
E	F	A	B	C	D
F	A	B	C	D	E

Figure 9.7 Systematic 6 × 6 Latin square to be used as starting point for randomization.

first the rows, as is shown in Figure 9.8, and then randomly to rearrange the columns, producing the randomized Latin square shown in Figure 9.9.

Analysis of the data collected in the experiment on amount of concentration of fertilizer, utilizing the Latin square design shown in Figure 9.6, follows the same logic as has been considered already in connection with the randomized blocks design, with one new complicating feature. This analysis is exemplified in the breakdown of *df* shown in Table 9.3.

Confounding in the Latin Square Design

Examination of the breakdown of *df* shown in Table 9.3 indicates that, after the removal of 5 *df* for rows and 5 *df* for columns, the *df* for error

D	E	F	A	B	C
B	C	D	E	F	A
F	A	B	C	D	E
A	B	C	D	E	F
C	D	E	F	A	B
E	F	A	B	C	D

Figure 9.8 Systematic 6 × 6 Latin square after random rearrangement of rows.

D	B	A	E	F	C
B	F	E	C	D	A
F	D	C	A	B	E
A	E	D	B	C	F
C	A	F	D	E	B
E	C	B	F	A	D

Figure 9.9 Randomized 6 × 6 Latin square, after rows and columns have been separately rearranged.

TABLE 9.3

Breakdown of *df* for analysis of variance of data from experiment on concentration of fertilizer employing a 6 x 6 Latin square.

Source	*df*
Concentrations (Latin letters)	5
Rows	5
Columns	5
Error	20
Total	35

can no longer be comprehended simply in terms of the concentrations x rows or the concentration x columns interactions, although this error term does, in fact, consist of these effects. The reason for this peculiarity of the Latin square design stems from the fact that a subtle and unavoidable form of confounding exists in this arrangement. The nature of this confounding may be understood by consideration of a simpler square in which it is quite easily identified. Consider, for this purpose, the 2 x 2 Latin square shown in Figure 9.10. It will be recalled that in our discussion of interaction in 2 x 2 factorial designs, in Chapter 8, it was pointed out that the interaction in a 2 x 2 design involves *a diagonal contrast* of half of the data with the other half.

In the 2 x 2 square shown in Figure 9.10, similarly, it is clear that the comparison of the units receiving the A treatment with those receiving

A	B
B	A

Figure 9.10 A 2 × 2 Latin square illustrating complete confounding of treatments with the row–column interaction.

the B treatment (the effect of Latin letters) is identical with the contrast defining the row x column interaction, since the two cells involving the A treatment are on one diagonal while the two cells receiving the B treatment are on the other diagonal. It can be demonstrated similarly that the rows x treatments interaction is identical with the columns effect and the columns x treatments interaction is identical with the rows effect. Each effect is defined, in fact, by the interaction of the remaining two factors in 2 x 2 Latin square design. This point may be clarified somewhat by observing that there are only 3 *df* available from the four cells in Table 9.10. If one of these *df* is attributed to rows and one to columns, there is only a single *df* left for the effect of the treatments on the face of the square. But this *df* is also the one defining the row x column interaction.

In the 2 x 2 Latin square, thus, the effect of treatments is completely confounded with the interaction between rows and columns. Even if the 2 x 2 square were replicated several times, it would be necessary to assume that no row x column interaction could exist in order to arrive at any conclusions regarding the effect of the treatments variable on the face of the square. For this reason, the 2 x 2 square is essentially useless in both agricultural and psychological experimentation. As the *order* of the square (i.e., its number of rows or columns) increases, however, the extent to which this type of confounding reduces our ability to draw conclusions regarding the independent variable on the face of the square quickly diminishes. The larger the square, the smaller the problem. But, even in very large squares (e.g., 8 x 8 or 10 x 10), the effect of the independent variable is still *partially* confounded by the row x column interaction. As a rule, therefore, the Latin square design is limited to experimental situations in which the possibility of a real row x column interaction effect is negligible on a priori grounds.

Repeated Measurements

The use of the Latin square in psychological experimentation has been confined almost exclusively to situations in which repeated measurements are made on the same subjects (i.e., subjects constitute one dimension of blocking) and it is necessary to control the order in which the subjects receive the various levels or categories of the independent variable. Although the topic of repeated measurements on the same subjects is taken up in detail in Chapter 10, it will be useful to consider this use of the Latin square design in the context of the foregoing presentation.

Consider, for example, an experiment on the effect of intralist stimulus similarity on rate of learning of lists of nonsense syllables. Suppose that it has been decided that six levels of intralist stimulus similarity will be employed in the experiment. Individual differences in rate of learning of lists of nonsense syllables are known to be quite large, so that it is necessary to use each subject as his own control. However, when each subject is observed at all levels of the independent variable, the *order* of occurrence of the different conditions must be controlled. It obviously would not do to permit any one of the levels of the independent variable to occur *first* all of the time, or *last* all of the time, or for that matter, in any ordinal position *all* of the time. This is because of the possibility that subjects may learn how to learn from list to list or because learning each list may interfere with learning subsequent lists, etc. It is necessary not only to control the ordinal position of occurrence of each level of the independent variable, so that each occurs in each ordinal position as often as the others, but also to give concern to the sequence (the entire set of six treatments given to a particular subject) in which the levels occur.

Both of these problems are effectively dealt with by the use of a 6 x 6 Latin square for this experiment. The rows of the square correspond to *sequences* of administration of the levels of the independent variable, while the columns of the square refer to *ordinal positions* (i.e., "first," "second"). Figure 9.11 illustrates a Latin square used in this way, with the letters on the face of the square referring to the different amounts of intralist stimulus similarity (the independent variable). The rows and columns of the square have been randomized.

Each of the levels of the independent variable appears once in the first ordinal position, once in the second, once in the third, etc. In the

ORDINAL POSITION

SEQUENCES (SUBJECTS)	1st	2nd	3rd	4th	5th	6th
S_1	E	B	D	A	F	C
S_2	C	E	F	B	A	D
S_3	B	A	C	E	D	F
S_4	F	D	B	C	E	A
S_5	A	F	E	D	C	B
S_6	D	C	A	F	B	E

Figure 9.11 Latin square design for conducting the experiment on the effect of intralist stimulus similarity on rate of learning of lists of nonsense syllables. Each row in the square requires at least one subject. The letters on the face of the square represent the six different levels of the independent variable.

context of this hypothetical experiment, the assumption that there is no interaction between rows and columns in the square can be quite explicitly understood. The assumption of no interaction requires that subjects who perform better (or poorer) than other subjects do so consistently, in all ordinal positions. Put another way, the assumption means that any benefit or detriment resulting from being first, second, or last in the sequence of lists to be learned will accrue to a list consistently, regardless of which subject's performance is considered.

On a priori grounds, it does not seem unreasonable to assume that the row–column interaction will be an appropriate estimate of error rather than arising from systematic tendencies toward real interaction. That is, if there is something in this experimental situation which engenders better performance early as opposed to late in the sequence of lists, or vice versa, it would appear that, whatever this influence might be, it would influence all of the subjects similarly.

Assumption of Zero Interaction

Certain writers (e.g., McNemar, 1955, p. 340), pointing to the fact that about half of the interactions reported in the experimental literature turn out to be significant, have argued that the assumption of zero interaction is unjustified and that this may render the Latin square design useless in psychology.

Two points are overlooked in this argument. First, it is likely that many experiments in which interactions could not be tested because no appropriate error terms were available were actually designed with the knowledge that interactions would not be present. In other words, the sample of experiments referred to by McNemar (in which half of the interactions were significant) was a group of experiments which were purposely designed to permit the testing of interactions, i.e., experiments on phenomena in which interactions were expected. What proportion of the total number of experiments done does this specialized sample represent?

Second, many of the interactions embraced by McNemar's generalization did not include *subjects* as one of the factors. The tendency to assume that each and every different human subject not only differs beforehand from every other human subject but is differentially influenced by independent variables is based in substantial part upon prejudice and confusion, since the two aspects of this tendency are more likely to be mutually incompatible than complementary. The very fact that people differ consistently is what makes their differences important in psychology. But, if they differ *consistently,* their differences may be found under a number of deliberately manipulated conditions. The latter point would suggest that interactions of manipulated independent variables and subjects may be *less* of a problem than McNemar claims.

Table 9.4 shows the breakdown of *df* for the analysis of variance of the Latin square experiment on intralist stimulus similarity, using a sample of six subjects, one in each sequence. Table 9.5 shows the breakdown of *df* when there are *n* subjects in each sequence, i.e., when the square is replicated *n* times (as would usually be done in an experiment of this type). In both of these *df* breakdowns it is assumed that subjects are assigned randomly to sequences and, in the replicated case, it is assumed that *the same* Latin square is used in each replication.

More Than One Square

In certain cases it may be necessary to use two or more different randomized Latin squares, with a single set of six subjects in each square (or as many subjects in each square as the *order* of the square requires). When more than a single square is used in the experiment, naturally, a larger number of different sequences of conditions are sampled. If there are a priori reasons for suspecting that certain patterns of successive conditions may have unusual idiosyncratic effects, it may be advantageous

TABLE 9.4

Breakdown of *df* for analysis of variance of Latin square design in study of intralist stimulus similarity. Six subjects are run in this study.*

Source	*df*	
Sequences (subjects)	5	$(r - 1)$
Ordinal positions	5	$(k - 1)$
Stimulus similarity	5	$(s - 1)$
Error	20	$(r - 1)(r - 2)$
Total	35	$(r^2 - 1)$

* Given in parentheses after each *df* is a general expression for determining *df* for squares of any order. The key to the symbols used in this table and in Table 8.5 is: r = number of rows, k = number of columns, s = number of levels of stimulus similarity, and n = number of subjects in each sequence.

TABLE 9.5

Breakdown of *df* for analysis of variance of Latin square design in study of intralist stimulus similarity, using *n* subjects in each sequence. The key to the symbols in this table is the same as in Table 9.4.

Source	*df*
Sequences	$r - 1$
Subjects/within sequences	$r(n - 1)$
Ordinal positions	$k - 1$
Stimulus similarity	$s - 1$
Error	$r(k - 1)(n - 1) + (r - 1)(r - 2)$
Total	$nr^2 - 1$

to use more than a single square. In most psychological applications of the Latin square design (as in the current example), however, ordinal position effects are more a matter of concern than are these more subtle sequence problems, so that the same square is replicated as a rule. The analysis shown in Table 9.5 has the advantage of permitting an appropriate test to be made of the effect of sequences, using the subjects/within sequences as the error term (i.e., sequences are replicated across subjects/within sequences).

Factorial Designs in Latin Squares

Just as was true in the randomized blocks design, factorial as well as single-factor experiments may be run in Latin squares, when subjects are to receive all of the combinations of conditions and the order of their occurrence is thought to be a potential problem. As in randomized blocks designs, the use of Latin squares for factorial experiments that are entirely within-subject is limited only by the size of the square that is feasible to employ. As will be shown in Chapter 11, a modification of the Latin square principle is available for those cases in which the number of combinations of conditions generated by a factorial design is too great to be used with any single subject (because of fatigue, etc.). Since this modified, or "quasi-" Latin square design involves the use of deliberate confounding, consideration of it will be postponed until Chapter 11.

GRECO-LATIN SQUARES

The simplest way to do the experiment on the effect of intralist stimulus similarity on rate of learning of lists of nonsense syllables would be to use the method of *serial anticipation.* The subject is shown the first nonsense syllable in the list and must attempt to identify the next stimulus in the series before it appears. If he fails to do so, of course, this is an error and the next syllable is presented to him. Now he must anticipate still another stimulus in the series. Eventually, the subject will be able simply to name all of the stimuli in the series without error. In a manner of speaking, the method of serial anticipation is one in which each item in the series serves as both a stimulus (for the subsequent syllable) and a response (to the preceding syllable).

Even though this method may be the easiest to employ, the experimenter may wish to study the effect of intralist stimulus similarity in a situation in which the stimuli *do not* also serve as responses. In this case, it would be necessary to use the method of *paired associates,* in which each stimulus is paired with its own, unique response. The stimulus is presented and the subject must correctly anticipate the response. If he fails, it is shown to him and the next pair is presented in the same way, etc., until the last response-syllable is shown. It is quite possible that the degree of intralist similarity of stimuli will affect rote learning differently (or in different degrees) when the stimuli are also responses as compared with when each stimulus has its own, unique response. Thus, it is likely that the paired associates method would have to be used to study this problem.

Now, when each stimulus-syllable has its own response-syllable, the lists will differ not only in manipulated amount of stimulus similarity (as the experimenter intends); they will also differ in that different sets of response-syllables are used with each list of stimulus-syllables. These response syllables may be unequally difficult in association with particular stimuli, even if careful attention is given to the task of selecting them to be generally equal in difficulty.

It is possible to conduct this experiment with different sets of response-syllables (along with the lists of stimulus-syllables which were deliberately constructed to differ in intralist similarity), in such a way that the conjoint occurrence of each particular stimulus list and each particular response list is balanced in the same way that lists were balanced in the more simple serial anticipation experiment. Each stimulus list will be presented with each response list once and only once. The ordering of response lists is controlled exactly as the ordering of stimulus lists was controlled previously, by another Latin square arrangement. Since no 6 x 6 Greco-Latin square can be constructed,[2] however, it will be necessary to shift to a 7 x 7 plan for illustrative purposes.

Two orthogonal Latin squares form a Greco-Latin square when superimposed. The term "orthogonal" here refers to the fact that each Latin letter appears once and only once with each Greek letter. Both alphabets are, of course, in a Latin square arrangement with rows and columns, such that each Latin letter appears once and only once in each row and in each column and the same thing is true for each Greek letter. If we let the Latin letters stand, as before, for amount of intralist stimulus similarity (the main independent variable) and let Greek letters stand for the different sets of response syllables, the experiment can be conducted as illustrated in Figure 9.12, utilizing a randomized Greco-Latin square design.

The procedure for constructing a randomized Greco-Latin square is similar to that shown for constructing a randomized Latin square, except that it is necessary to start with a systematic Greco-Latin square and rearrange the rows and columns using the *pairs* of Greek and Latin letters, taken together; then, separately to rearrange randomly either the Latin or Greek rows while holding the other alphabet fixed.

The analysis of data collected using the Greco-Latin square design shown in Figure 9.12 involves no new principles. For illustrative purposes, Table 9.6 presents the breakdown of *df* for this analysis, assuming that one subject is assigned to each sequence.

[2] Greco-Latin squares can be constructed for any prime number or square of any prime number above 2.

ORDINAL POSITION

SEQUENCES (SUBJECTS)	1st	2nd	3rd	4th	5th	6th	7th
S_1	C α	E ϵ	D γ	F η	A δ	G β	B ζ
S_2	A ζ	C γ	B α	D ϵ	F β	E η	G δ
S_3	B η	D δ	C β	E ζ	G γ	F α	A ϵ
S_4	G ϵ	B β	A η	C δ	E α	D ζ	F γ
S_5	E γ	G η	F ϵ	A β	C ζ	B δ	D α
S_6	F δ	A α	G ζ	B γ	D η	C ϵ	E β
S_7	D β	F ζ	E δ	G α	B ϵ	A γ	C η

Figure 9.12 Randomized Greco-Latin square for experiment of intralist stimulus similarity using different response lists.

TABLE 9.6

Breakdown of *df* for analysis of variance of Greco-Latin square design shown in Figure 9.12, assuming one subject in each sequence.

Source	*df*
Sequences (subjects)	6
Ordinal positions	6
Stimulus similarity (Latin)	6
Response lists (Greek)	6
Error	24
Total	48

MATCHED PAIRS DESIGNS IN PSYCHOLOGICAL RESEARCH

Aside from the use of subjects as their own controls (which is dealt with separately in Chapter 10), the next most frequent use of the blocking principle in psychological experimentation is probably to be found in studies in which subjects are run in pairs, matched on some psychometrically defined variable that is assumed to be correlated with the dependent variable.

When used appropriately, the matched pairs design is a highly effective device for reducing error variance arising from individual differences in the behavior to be observed. Although this method is restricted to situations in which the independent variable has only *two* levels or categories, experiments of this type are very common in psychological research, especially in light of the prevalent emphasis on simple hypothesis testing that characterizes much of it.

The effectiveness of the matched pairs design stems from the fact that the block, consisting as it does of only two subjects, is capable of maximum homogeneity relative to the heterogeneity of the entire sample of subjects. It is obvious that intrablock homogeneity is inversely related to block size, all other things equal. It is much easier to assemble *pairs* of subjects who are very similar in some measure that is correlated with the dependent variable than it is to form trios, quartets, or larger groups of similar subjects. Since the object of blocking is to permit the experiment to be conducted within experimental units that are highly similar, it is easy to see that *pairs* meet this requirement more readily than would groups of any larger size.

Reduction of Error in Matched Pairs

The extent to which reduction of error is achieved by blocking the subjects into matched pairs is elucidated by noting that the standard error of the difference between two means (the denominator of the appropriate t test in matched pairs experiments) is reduced to the relative extent that the dependent scores of the members of the matched pairs are correlated with each other. This relationship is illustrated by the formula for the standard deviation of the differences between two sets of scores,

$$\sigma_{1-2} = \sqrt{\sigma^2 + \sigma^2 - 2r_{12}\,\sigma_1\,\sigma_2},$$

in which it should be clear that *perfectly* correlated scores would result in a standard deviation of zero, since, in that case, σ_1 would equal σ_2 and the terms on either side of the minus sign would be identical.

Some misunderstanding of the foregoing relationship is apparent in the literature (e.g., Edwards, 1960, p. 172), since it is sometimes pointed out that the standard error of the difference between two means *may be inflated* if the intercorrelation between the paired scores is negative. While this observation is perfectly accurate, in the light of the minus sign in the formula, it is entirely trivial, since careful consideration of the actual procedure that is used to achieve matching in the first place should make it clear that a negative correlation between the scores of matched subjects would almost never occur, even when the variable on the basis of which matching is achieved is negatively correlated with the dependent variable.

Procedure for Assembling Pairs

Matching is accomplished by rank-ordering the subjects on the matching variable (presumably one that is substantially correlated with the dependent variable) and then segregating them into pairs based on this ranking (i.e., 1 and 2 in the first pair, 3 and 4 in the second, etc.). The two members of each pair are likely to be closer to one another in the measured response than either is to any other subjects in the group. Hopefully, a wide range of scores has been obtained, since the more variable the subjects are before matching, the more there is to be gained from matching.

Following the standard procedure of the randomized blocks design after pairs are formed, the subjects within the pairs are assigned at random to the two conditions. It does not matter whether the correlation between the matching and the dependent variables is negative or positive, as long as it is substantial (above 0.50). In either case, pairs will consist of subjects who performed similarly on the matching variable. If the correlation between it and the dependent variable is substantial, it goes without saying that the performance of members of a pair on the dependent variable will be *positively* correlated.

A frequent misguided modification of the matched pairs design is used in psychological experimentation, in which two *groups* of subjects are assembled such that they have equal means and variances on some measure known or thought to be correlated with the dependent variable. This method, sometimes called the matched groups design, typically involves some juggling of subjects back and forth from group to group, until the desired equality of means and variances is achieved.

Even when the matching variable is highly correlated with the dependent variable, this design is ineffective in reducing error variance due to individual differences, since it provides no method for partitioning the between-subject variance into its within-pair and between-pair components. This failing is due to the fact that the subjects were not actually *paired* and randomly assigned to the two conditions within pairs. In fact, as has been noted earlier in this chapter, any procedure which tends to reduce the observed differences between the two groups' means (as this method does) will result in a *loss* in sensitivity if an appropriate reduction of the error variance cannot be effectuated.

A great many experiments have been conducted under these manifestly weakened conditions, since many experimenters strive to "match" the subjects of their experimental groups on any number of "relevant" variables (e.g., educational level, chronological age, mental age, race, socio-economic level, etc.), operating on the mistaken assumption that this type of matching is inherently desirable. Any technique for increasing the homogeneity of the experimental material for an experimental contrast without providing for an appropriate reduction of the error variance will have the effect of increasing the probability of a Type 2 error. In addition, even when appropriate procedures are provided for, only variables which correlate substantially with the dependent variable (over 0.50) will achieve the desired error-reduction anyway.

YOKED CONTROL DESIGNS

In recent years, a modification of the matched pairs design has come into vogue which blocks the subjects (in pairs) in terms of concomitants of the experimental procedure other than those thought to be "crucial" to the experimental question. For example, in studies of "the executive monkey," subjects are run in pairs such that they both receive the paired stimuli in identical sequences, except that the "experimental" subject's response terminates or prevents the electric shock for both animals.

Confounding in Yoked Control Designs

As Church (1964) has observed, these studies may involve a type of inadvertent confounding which is prejudicial to the contrast of interest, if the subjects within pairs differ in sensitivity to the various conditions of the experiment. For example, if one member of a pair is a better learner than the other and he happens to be the "experimental" subject, his behavior will prevent the poorer learner from receiving a sufficient number of trials of paired stimuli to enable him to learn. However, when the poorer learner is the "experimental" subject, this bias does not cancel out, because the poorer learner will require many paired trials to achieve a level of responding that will prevent electric shocks, but the better learner will not benefit from these extra pairings. This asymmetry will result in a tendency for better performance in the "experimental" condition, even if no actual psychological effect beyond the sampling differences mentioned is operating.

An additional problem of the yoked control design, in the present context, is the frequently overlooked need to adjust the error variance to allow for the blocking that results from the yoked control procedure.

An illuminating example of the dangers of yoked control designs is found in a study by Gormezano, Moore, and Deaux (1962). In this study, subjects were run in yoked pairs in a comparison of classical and avoidance eyelid conditioning under three levels of intensity of the unconditioned stimulus. Both members of each yoked pair received the airpuff only on those trials when the avoidance member made a criterion eyeblink. A positive relationship between level of conditioning and intensity of airpuff was found in the three groups of avoidance subjects, but the relationship was negative in the three control (classical) groups. The explanation of this result was to be found in the fact that the more responsive each avoidance subject was, the fewer the number of airpuffs his yoked control received and, thus, the lower the level of conditioned responding he achieved. In this way, any independent variable which increased the level of responding of the avoidance subjects necessarily reduced the level of responding of the classical controls, even if the eyeblink response cannot really be conditioned instrumentally. Naturally, the intensity of the airpuff was an independent variable of this type.

The importance of this analysis is that it shows that a difference in the influence of a major independent variable (unconditioned stimulus intensity) may be due entirely to the procedure of matching rather than any differential influence of the variable on the other possible features of difference between the two conditions (i.e., avoidance versus classical). While it is true that yoked control designs may have special virtues in certain types of experiments, only careful attention to their possible deleterious side effects can ensure that they are used effectively. Sometimes (Kimmel & Terrant, 1968) it may be possible to employ elaborations of the yoked control design to overcome confounding problems. Even when they are both especially appropriate and cautiously employed, appropriate adjustment of the error variance must also be accomplished to ensure that an accurate assessment of significance will be possible.

SUMMARY

The experimental materials of psychological experiments, human and subhuman subjects, display wide variations in the dependent behavior even when they are treated similarly. This *heterogeneity* of the experimental material may sometimes be dealt with effectively by application of the *blocking principle,* by means of which the basic comparisons of the experiment are made within homogeneous blocks of subjects. The simplest use of blocking is exemplified by the *randomized blocks design,* which was developed in agricultural research to deal with the related problem of soil heterogeneity. In randomized blocks experiments, the subjects are grouped according to a control variable known to be correlated with the dependent variable and, within these blocks, they are assigned at random to the levels of the independent variable. The size of the block is the same as the size of a replication. The appropriate error term for evaluating the effect of the independent variable is the *treatments-by-blocks interaction,* since it arises from replication of the phenomenon under study. Factorial designs may also be run in randomized blocks with few additional complications.

The *Latin square design* involves simultaneous double-blocking with respect to two control variables. In psychological experimentation, Latin squares have been used most often to control *order effects* at the same time that *subject differences* are controlled by blocking completely within-subjects. Because of the inherent partial *confounding* of the row–column interaction with the treatment effects in Latin squares, special consideration of this possibility must be given. Factorial designs may also be run in Latin squares, again with little in the way of added complexity. If it becomes necessary to control still another extraneous factor in a completely within-subject experiment to be run in Latin squares, a *Greco-Latin square design* may be employed. In the area of verbal learning, it is not uncommon for such methods to be used with beneficial results.

The *matched pairs design* is a special instance of the blocking principle which has been widely used in psychological research. Subjects are placed in rank order on some control variable which is known to be highly correlated with the dependent variable and are then separated into pairs (the highest two, the next two, etc.). Within pairs, subjects are assigned at random to the two conditions. The error variance will be reduced to the extent to which the control-dependent variable correlation accounts for substantial variation.

When used effectively, matched pairs may increase the sensitivity of an experiment considerably. However, matching groups of subjects in mean and variance, without providing for appropriate reduction of error as indicated above, will often result in *less sensitive* studies. A recent variation of the matched pairs design, involving *yoked controls,* permits subjects to receive the same stimulus conditions while differing with respect to response contingencies. Although these designs may involve special confounding problems, their advantages are testified to by the rapid increase in frequency with which they are reported.

10

Blocking by Repeated Measurements

As was indicated at the beginning of Chapter 9, heterogeneity of the experimental material would be dealt with ideally if, hypothetically, we could measure the same individual's behavior under all of the different experimental conditions *simultaneously*. The same experimental subject would provide a behavior measure at each of the different levels of the independent variable. Comparisons among these levels would enable the effect of the independent variable to be assessed completely "within-subject."

More than a single subject would be needed only to provide a basis for generalizing to some population of subjects. In a situation of this type, when reference is made to a "random" sample of subjects, the word "random" refers both to the possible overall differences in level of performance among the subjects as well as to differences among subjects in their unique patterns of differential response to the various levels of the independent variable (i.e., both subject differences *and* the treatments x subjects interaction are randomly sampled). It should be clear that a *hypothetical*, completely within-subject experiment of this type could achieve maximum sensitivity to the effects of the independent variables.

Even though it is patently impossible "simultaneously" to measure an individual's behavior under two or more different experimental conditions, a great many experiments are conducted either partly or fully within subjects, simply on the basis of the assumption that making *repeated* observations on the same subjects at different times is almost the same as hypothetically making them all at one time. There are many experimental situations in which this assumption appears to be quite justified.

For example, suppose that an experiment is to be conducted on the effect of social competition on the performance of a highly trained task. The task to be employed is the running of a footrace. High school seniors are to serve as the subjects and it is assumed that they have run over the selected course many times previously. The independent variable is defined by the different competitive conditions under which the race is to be run:

1. racing alone, "against the clock";
2. racing against a low-prestige classmate;
3. racing against a classmate of average prestige;
4. racing against a high-prestige classmate.

The dependent variable is the time taken by an individual to run the race.

It is highly likely that interindividual variations in running time would constitute a serious source of heterogeneity in an experiment of this type and would require that a very large number of subjects be used if a random groups design were selected. The members of a typical high school senior class are likely to differ greatly in their ability to run a footrace under standard conditions, even assuming that they have all had extensive and approximately equal practice. Blocking the subjects into relatively homogeneous ability groups on the basis of a preliminary time measure (or by age or body weight, either of which may be correlated with running ability) would permit some reduction in the overall size of the study, because of the increased sensitivity of the randomized blocks design. But, using repeated measurements *within* subjects provides the ideal method for dealing with this particular heterogeneity problem, since it is easy to arrange for rest periods between races (if necessary, this could be carried to the point of conducting the experiment on four different days). With adequate rest periods between trials, it may be assumed that the subject is in more or less the same state of readiness to perform for each of the four trials.

To further ensure that the subject is theoretically "the same" under each condition, certain steps might have to be taken to avoid the possibility that feedback on any one trial (i.e., information regarding how well or poorly he performed) might influence performance on subsequent trials. In the present instance, for example, it might be necessary to have the contestants run the track in opposite directions, so that the subject would not discover who had "won" the race. Or, some kind of visual barrier might be used to separate the runners, etc.[1]

Progressive Errors

In general, the kinds of factors that must be considered to be potentially deleterious in experiments involving blocking within subjects are those that may result in systematic decrements in performance from trial to trial (e.g., *fatigue,* which is dealt with in the present example by allowing for an adequate amount of time for rest), those that may result in systematic increments from trial to trial (e.g., *learning,* or practice effects, which is dealt with in the present example by employing an already highly practiced task), and those that may result from informational feedback (or other trial-specific long-term changes) regarding performance on any given trial, except the last. To the extent to which none of these presents an insurmountable obstacle to the experimenter, within-subject blocking may be used to considerable advantage.

TREATMENTS X SUBJECTS DESIGN

When the individual subject serves as the block within which all of the levels of the independent variable are to be studied, as in the footrace example discussed above, the experiment exemplifies what is frequently referred to as a treatments x subjects design. Each subject, in these cases, is said to serve as "his own control," in that his performance at each level of the independent variable is compared against his own average performance. Just as is true in the parent randomized blocks design, differences between blocks (in this case, between subjects) are eliminated as a source of error, since intersubject variability does not contribute to the appropriate error estimate.

In the randomized blocks design it is necessary to assign the different levels of the independent variable randomly to subjects within the blocks; in the treatments x subjects design, accordingly, it is necessary to randomize the *order* of delivery of the different levels of the independent

[1] As will be discussed more fully in Chapter 12, it may sometimes develop that apparently necessary control procedures have the effect of "weakening" the independent variable. In the present example, any attempt to reduce the informational feedback may also reduce the effective differences in social competition; it certainly would matter less against whom one is racing if the opponent cannot even be seen. Obviously, if "feedback" is an essential aspect of the definition of "competition," it might be necessary to use a randomized blocks design for this experiment, since the presence of feedback would contaminate the social competition variable (i.e., the independent variable would become a combination of competition plus feedback).

variable within each subject. Otherwise the two designs are identical in structure and in analysis. Table 10.1, which shows the breakdown of *df* for analysis of variance of data collected in a treatments x subjects experiment, illustrates this point, with subjects substituting for blocks and with the treatments x subjects interaction substituting for treatments x blocks as an estimate of error.

TABLE 10.1

Breakdown of *df* for treatments x subjects design.

Source	*df*
Subjects	$n - 1$
Treatments	$a - 1$
Error (treatments x subjects)	$(a - 1)(n - 1)$

Appropriate Error Estimate

That the treatments x subjects interaction provides an appropriate estimate of error for evaluating the significance of the effect of the independent variable (i.e., treatments) should be clear from the previous discussion of the randomized blocks design. The "phenomenon" under study in this case, of course, is the pattern of differences associated with the different levels or categories of the independent variable. Replication across subjects results in repeated observations of this "phenomenon." The extent to which the "phenomenon" varies from subject to subject is exactly what is manifested in the treatments x subjects interaction. The more consistently is the "phenomenon" reflected in the data from subject to subject, the more confident may the experimenter be that it is not due to chance.

Still another approach to establishing that the treatments x subjects interaction is the appropriate source of error for testing the effect of treatments may be found in the previous observation that random sampling of subjects results in random sampling of their unique patterns of reactions to the different levels of the independent variable. As was noted in Chapter 7 (pp. 142–43), the expected value of the variance estimate for the *fixed* factor (i.e., treatments) in experiments based upon the mixed model for analysis of variance (ideally exemplified in the treatments x subjects design in psychology) contains a component of variation due to the randomly sampled interaction, along with a component due to the treatments factor and another due to random error. The expected value of the variance estimate for the treatments x subjects interaction consists of a component due to the interaction and another due to random error. By the rule of thumb that the denominator of the appropriate F ratio should contain all of the components of the numerator except the effect being tested, the treatments x subjects interaction should be used to test the effect of treatments. Whether or not there is a real interaction, such an F test will indicate whether the treatments effect is significant.

Testing the Significance of the Interaction

No provision is made in the treatments x subjects design considered above for the possibility that the experimenter may be explicitly interested in the interaction itself. Since replication of observation of the interaction is not accomplished, no error term exists for testing its significance (i.e., the total sum of squares is exhausted by successively subtracting the sums of squares for subjects, treatments, and finally, treatments x subjects). Although subjects are measured repeatedly, each subject is measured only once at each level of the independent variable.

The point being discussed, in fact, is a very good illustration of what is meant by the statement that provision must be made for replication of the "phenomenon" of interest, since without such replication there is no source of variability which may be used to test the significance of the "phenomenon." In order to provide an appropriate estimate of error for testing the significance of the interaction in the treatments x subjects design, each subject must be measured at least twice at each level of the independent variable (and, hopefully, these two or more observations should differ only randomly).

In the context of the footrace experiment discussed above, the experimenter may have perfectly sound psychological reasons for wishing to determine whether the effect of the social competition factor differs significantly among his subjects. That is, he may wish to know whether the lack of uniformity in the effect, from subject to subject, is itself a stable feature of the population from which the subjects were obtained. To ensure that he will be able to answer this question, he must arrange to replicate his measurement within each subject at least twice at each level of the competition variable.

In the footrace experiment this would not necessarily require that the same low-prestige peer be employed more than once, for example, since two randomly different low-prestige peers would serve the same purpose. Assuming that there are two within-subject replications at each level of the competition variable, there would then be eight different measurements made on each subject, two at each of the four levels of competition. The order of delivery of these eight treatments must be separately randomized for each subject, again with adequate precautions being taken to provide adequate rest periods, etc. Table 10.2 shows the breakdown of *df* for analysis of variance of these data, the letter "*r*" being used for the number of within-subject replications.

TABLE 10.2

Breakdown of *df* for treatments x subjects design with *r* replications within subject.

Source	*df*
Subjects	$n - 1$
→Treatments	$a - 1$
→Treatments x subjects	$(a - 1)(n - 1)$
Replication within Ss	$an\ (r - 1)$

The effect of treatments, as in the simpler version of this design, is tested with the treatments x subjects interaction as the error term. The treatments x subjects interaction is tested using replication within subjects as an estimate of error. Both of these tests are indicated by the arrows in the table which start at the denominator and point to the numerator. For computational purposes, the latter error term may be obtained as a residual, by subtracting the other sums of squares from the total sum of squares.

Although the treatments x subjects interaction is ordinarily used to test the effect of treatments, if the experimenter wished to determine whether the treatments factor had any effect whatever, *either* separately *or* in interaction with subjects, he could use replication within subjects (the within-cells variance estimate of previous examples) as an error estimate in the denominator of an *F* ratio testing the treatments effect. But, if this *F* were statistically significant, he would be confined to a conclusion of the all-inclusive type just verabilzed, i.e., the social competition factor has an effect on racing speed (in the language of the foot-race experiment), *either* because competition conditions influence running speed differentially, in general, or because competition conditions differentially influence running speeds of some subjects differently from others. Since this design permits separate tests to be made of the two separate components of this broad conclusion, it would seldom be desirable to resort to this type of all-inclusive test. This brief discussion of it is intended primarily to illustrate the principle that different estimates of error are appropriate to test different types of hypotheses and that each error estimate may require its own type of replication procedure.

Simultaneous Repeated Measurements

Frequently, it may actually be possible to arrange the experimental task so that the subject's behavior is in effect being measured under all of the levels of the independent variable at more or less "the same time." For example, in an experiment dealing with the effects of stimulus factors on perceptual responses or upon simple learning, it is not uncommon for each "trial" to contain instances of each level of the independent variable, the order of their occurrence being separately randomized for each trial. The dependent variable may be the sum of several trial-scores, so that it is almost as if all of the levels of the independent variable *were* really administered simultaneously.

Zeigarnik's famous experiment (1927) on memory for completed and interrupted tasks was of this form, since each subject attempted all of the twenty tasks employed, one after another. Half of the tasks were completed and half were interrupted prior to completion. The order of administration of the two types of tasks (completed and interrupted) was unsystematic and more or less balanced across the twenty trials. After the last task, the subject was instructed to write down the names of as many of the tasks as he could recall. Two scores were obtained, the proportion of completed tasks recalled and the proportion of interrupted tasks recalled. The manner in which the two "treatments" were administered within subject exemplifies what is meant by "almost simultaneous" administration of the different conditions, since neither type of task could be said to have been given first or second in relation to the other, either during their actual presentation or when the actual scores were obtained.

To be sure, this experiment is flawed by the possibility that only a limited memory storage may be available for short-term recall, so that a real tendency for better recall of one of the two types of task might have the effect of attenuating recall scores on the more poorly remembered task. If a person could only remember, say, about ten tasks, and if there were a slightly greater tendency to remember interrupted tasks, these would "fill up" the storage and leave no room for the completed tasks. This nonindependence of the two scores might result in an exaggeration of the actual size of the effect in an experiment of this type.

This type of nonindependence is different from that which results from the fact that repeated performances by the same subject are likely to be correlated with one another. The latter type of nonindependence is an essential aspect of all designs involving repeated measurement within subjects and is in no way a deficiency.

Repetition Over Trials

A special problem arises in many experiments utilizing within-subject blocking to increase sensitivity in the test of one or more independent variables, when one or more of the factors whose levels are administered completely within subjects represents the successive stages of some kind of sequential process, such as conditioning or extinction *trials* in learning studies, or successive *days* in other types of longitudinal studies. These applications of the blocking principle deviate somewhat from the fundamental notion that randomization must be employed within blocks, since it is manifestly impossible to randomize the order in which successive trials or days occur. Trial number 1 must always be administered first and trial number 2 must always be second, etc. Although this discrepancy has hardly had the effect of dissuading experimenters from using the logic of randomized blocks in situations of this type, as will be exemplified by the number of quite common varieties discussed below, it is well to recognize the possibility that underestimates of the appropriate error terms may result from this modification of the blocking principle.

FACTORIAL EXPERIMENTS WITHIN SUBJECTS

When two or more independent variables are each capable of administration completely within subjects, a design which is sometimes labeled treatments x treatments x subjects (or *A* x *B* x subjects) may be employed. No new principles are involved beyond those already considered in connection with factorial designs in randomized blocks and in the simple treatments x subjects experiment. While in the randomized blocks application it is conventional simply to pool the several treatments x blocks interactions into a single error term, the necessarily assumed homogeneity of the separate treatments x subjects interactions is less often justified in the within-subjects situation. Thus, it is often necessary to employ different error terms to test each separate treatment effect.

With reference to the breakdown of *df* for the analysis of variance of a factorial experiment completely within subjects, which is shown in Table 10.3, the effect of *A* is tested with *A* x subjects as the error term, *B* is tested with *B* x subjects, and the *A* x *B* interaction is tested with *A* x *B* x subjects as the error term. Arrows are again used to identify the appropriate terms of each *F* ratio.

For example, in an experiment on habituation of the orienting reflex to a simple visual stimulus (Kimmel & Kimmel, 1965), each subject received a sequence of repeated presentations of the stimulus each day

TABLE 10.3

Breakdown of *df* for factorial experiment completely within subjects.

Source	*df*
Subjects	$n - 1$
┌→ A	$a - 1$
└→ A x subjects	$(a - 1)(n - 1)$
┌→ B	$b - 1$
└→ B x subjects	$(b - 1)(n - 1)$
┌→ A x B	$(a - 1)(b - 1)$
└→ A x B x subjects	$(a - 1)(b - 1)(n - 1)$

for several days. The principle independent variables in an analysis were both within-subjects effects, trials (within days), and days. It was expected that the size of the orienting reaction would decline from the first to the last trial each day, would recover from the end of each day to the beginning of the next, and would decline overall across days. The last expectation, if confirmed, would also suggest that there should be a reduction in the amount of within-day reduction from day to day, since recovery at the beginning of each day would be to a lower and lower point.

The separate treatments x subjects interactions differed far too much to have justified pooling of them into a single error term. Thus, the effect of trials was tested with the trials x subjects interaction as the error term, days was tested with the days x subjects interaction, and the trials x days interaction was tested with the trials x days x subjects interaction. This study is also an illustration of the situation noted above, in which neither trials nor days are capable of being delivered in a random order within subjects. Thus, the significance of these effects may have been slightly overestimated.

WITHIN-SUBJECT EXPERIMENTS INVOLVING ORDER EFFECTS

In the within-subjects applications of the blocking principle considered so far (except those in which the independent variable is sequential and cannot be randomized—for which the following is irrelevant), it has been assumed that the *order* in which the different experimental conditions are administered is not a significant experimental factor in its own right. In such cases randomization of the order of delivery of the different treatments provides a basis for testing the null hypothesis regarding the effects of the treatments factor.

In within-subjects experiments involving independent variables which do not permit easy dismissal of the possibility of real order or carryover effects, as well as in those studies in which an explicit interest in order effects is present, it may be necessary to employ a Latin square design, as described in Chapter 9. In the Latin square design, blocking is employed to control for individual differences among subjects (in the rows of the square) and to control for ordinal position in the sequence (in the columns of the square). The Latin square may also be used in those cases in which the factorial combination of conditions from two or more independent variables are all to be administered completely within subjects and explicit concern exists regarding order effects: The only practical limitation upon the use of the Latin square design to control within-subject order effects is that the number of separate occasions on which each subject is to be measured must not become so large as to be unwieldy.

When it is clearly desirable to employ within-subject blocking in a factorial design involving potential order effects, but it is not feasible to administer all of the treatment-combinations to the same subject, a modified Latin square design which involves deliberate confounding of one of the factorial interactions (i.e., deliberate sacrifice of information) may be advantageous. Quasi-Latin squares of this type are described in Chapter 11 in connection with the general topic of deliberate confounding. Partial Latin squares (Youden squares) are also of possible use in cases of this type (Cochran & Cox, 1957).

TREND ANALYSIS IN THE TREATMENTS X SUBJECTS DESIGN

When the within-subjects independent variable is administered in equal quantitative steps, the sum of squares for treatments may be partitioned into orthogonal trend components in the same manner as was illustrated in Chapter 8 for single-factor random groups designs. Since the error term in the within-subjects design is the treatments x subjects interaction, however, this too must be partitioned into separate component error terms for testing the separate components of the trend (Grant, 1956).

Partitioning the Interaction

The treatments x subjects interaction, as has been noted previously, arises from the fact that different subjects may have different *patterns* of reaction to the various levels of the independent variable. These patterns may differ linearly, quadratically, etc. For example, in a hypothetical experiment on problem solving as a function of group size, the same subjects performed repeatedly, either alone, in two-man groups, in three-man groups, in four-man groups, or in five-man groups. On each occasion, the subject responded as an individual, following group interaction regarding the problem, and the time taken to arrive at the solution was recorded. Because it was anticipated that performance would benefit from group interaction only up to some optimum group size, after which the time taken to achieve problem solution was likely to increase, the overall F for group size was of less interest than was the quadratic component of the trend of time scores across group size. In other words, a U-shaped function was expected.

Computational Example

To illustrate the computations necessary for performing this trend analysis, Table 10.4 presents the results of ten subjects in this experiment. Although the group size factor is represented in the columns of Table 10.4 in an ordinal sequence, the actual order in which the different conditions were administered was separately randomized within each sub-

TABLE 10.4

Data (in seconds) from hypothetical experiment on group size in problem solving, illustrating computations for trend analysis.

	Group Size 1	2	3	4	5	ΣkX Linear	Quadratic
Linear coefficients	−2	−1	0	1	2		
Quadratic coefficients	2	−1	−2	−1	2		
Subject							
1	24	16	14	18	20	− 6	26
2	31	26	20	18	21	−28	20
3	18	12	13	15	18	3	19
4	27	24	19	18	18	−24	10
5	19	14	14	15	16	− 5	13
6	20	15	15	17	21	4	20
7	22	15	16	14	18	− 9	19
8	28	25	23	18	21	−21	9
9	19	18	15	15	15	−11	5
10	20	19	20	18	22	3	7
ΣX (T)	228	184	169	166	190	−94	148
$\bar{X}$	22.8	18.4	16.9	16.6	19.0		

ject. Since no explicit interest or possible psychological meaning existed for the higher-order components of the trend (i.e., the cubic and quartic), the table shows the orthogonal coefficients only for the linear and quadratic components.

The sum of squares for the linear component of the overall trend is obtained, as in the previous example of trend analysis, by entering the appropriate values into the formula:

$$\frac{(\Sigma kT)^2}{n\Sigma k^2}$$

Substituting the value (−94) for ΣkT (i.e., the sum of −2 times 228, −1 times 184, 0 times 169, 1 times 166, and 2 times 190); (10) for Σk^2; and with $n = 10$, this becomes:

$$\begin{aligned} SS_{\text{linear}} &= (\Sigma kT)^2/n\Sigma k^2 \\ &= (-94)^2/(10)(10) \\ &= 88.36 \end{aligned}$$

Similarly, the sum of squares for the quadratic component of the overall trend utilizes a value of (148) for ΣkT (i.e., the sum of 2 times 228, −1 times 184, −2 times 169, −1 times 166, and 2 times 190); (14) for Σk^2; and n again = 10:

$$\begin{aligned} SS_{\text{quadratic}} &= (\Sigma kT)^2/n\Sigma k^2 \\ &= (148)^2/(10)(14) \\ &= 156.46 \end{aligned}$$

The linear component of the treatments x subjects interaction is obtained by substituting appropriate values from Table 10.4 into the formula: [2]

$$\sum_1^n(\sum_1^a kX)^2/\Sigma k^2 - (\sum_1^n\sum_1^a kX)^2/n\Sigma k^2$$

In this case the separate stage scores for each subject are multiplied by the orthogonal coefficients to produce a

$$\sum_1^a kX_{\text{lin}}$$

for each subject. These are then squared and summed. Thus, for subject number 1,

$$\sum_1^a kX_{\text{lin}} = -2(24) + (-1)(16) + 0(14) + 1(18) + 2(20) = -6,$$

[2] $(\sum_1^n\sum_1^a kX)^2$ is, of course, the same as $(\Sigma kT)^2$ of the previous formulas.

After the

$$\sum_{1}^{a} kX_{\text{lin}}$$

have been squared and summed across the n subjects, the total is divided by Σk^2 and then

$$(\sum_{1}^{n}\sum_{1}^{a} kX)^2/n\Sigma k^2$$

is subtracted from this value to produce the resulting sum of squares for the linear component of the treatments x subjects interaction:

$$\begin{aligned} SS_{\text{treatments x subjects (linear)}} &= 2{,}098/10 - 88.36 \\ &= 121.44 \end{aligned}$$

The quadratic component of the treatments x subjects interaction is obtained in the same way as above, with the exception that the quadratic set of coefficients is used in computing

$$\sum_{1}^{a} kX \text{ and } \Sigma k^2.$$

Thus, for subject number 1,

$$\sum_{1}^{a} kX_{\text{quad}} = 2(24) + (-1)(16) + (-2)(14) + (-1)(18) + 2(20) = 26,$$

After

$$\sum_{1}^{a} kX_{\text{quad}}$$

have been squared and summed across the n subjects, the total is divided by Σk^2 and

$$(\sum_{1}^{n}\sum_{1}^{a} kX)^2/n\Sigma k^2$$

is subtracted to produce the resulting sum of squares for the quadratic component of the treatments x subjects interaction:

$$\begin{aligned} SS_{\text{treatments x subjects (quadratic)}} &= 2{,}622/14 - 156.46 \\ &= 30.82 \end{aligned}$$

For the purpose of providing a check on the accuracy of these computations, the cubic and quartic components of the treatments x subjects interaction may be computed in the same fashion as above (using the appropriate cubic and quartic coefficients), and the sum of the four separate components of the treatments x subjects interaction can be compared to the T x Ss sum of squares computed in the orthodox fashion.[3] Each of the separate components of the treatments x subjects interaction has 9 x 1 = 9 *df* and their sum, of course, contains the 9 x 4 = 36 *df* of the overall treatments x subjects interaction.

The *F* test of the significance of the linear component of the overall trend requires that the variance estimate for the linear component be divided by the variance estimate for the linear component of the treatments x subjects interaction:

$$F_{\text{lin}} = \frac{88.36/1}{121.44/9} = 6.55$$

Similarly, the *F* test of the significance of the quadratic component of the overall trend involves dividing the variance estimate for the quadratic component of the trend by the variance estimate for the quadratic component of the treatments x subjects interaction:

$$F_{\text{quad}} = \frac{156.46/1}{30.82/9} = 39.91$$

Each of these *F* ratios has 1 and 9 *df*, so that the *F* for the linear component is significant at $p < 0.05$ and the *F* for the quadratic component is significant at $p < 0.001$. On the basis of these statistical tests it would be concluded that speed of problem solving increases as a function of group size and that an optimum group size (somewhere around three or four) exists, above which speed tends to reduce.

FACTORIAL DESIGNS BETWEEN AND WITHIN SUBJECTS

Examination of the published research literature testifies to the fact that the most common use of the blocking principle in psychology occurs in factorial experiments in which it is possible for one or more of the independent variables to be administered within subjects while one or more others must be administered to different groups of subjects. In ex-

3 This same method can be used to check the accuracy of computation of the sums of squares of the linear and quadratic components of the overall trend, since the four separate single *df* components of the overall trend sum to the sum of squares for treatments.

periments of this type, sometimes called *mixed* factorial designs, tests of the within-subjects effects benefit from the increased sensitivity resulting from using subjects as their own controls. Tests of the between-subjects effects are no different in sensitivity from those in any other random groups design.

One Factor Between and One Within Subjects

The most common representative of the broad class of mixed factorial experiments is the simplest, with a single independent variable (usually, but not always manipulated), administered to randomly constituted groups of subjects, and another independent variable involving repeated measurements on the same subjects. Since, as has already been noted, successive trials are often dealt with as a within-subjects independent variable, any ordinary random groups experiment in which the dependent variable is measured on several occasions within the same subject is capable of being treated as a mixed factorial design. This is done most commonly in studies of learning and extinction. As will be explained below, this analytic device frequently provides the experimenter with a "second chance" to evaluate his general experimental hypothesis (i.e., that the independent variable has *some* effect), an advantage which has probably contributed to its popularity.

In a mixed factorial design with one independent variable manipulated between and one within subjects, the subjects are assigned randomly to b different groups, with n subjects in each group. Each group receives a different level of one of the independent variables, B. Within each of the b groups, each subject receives all of the a levels of the other independent variable, A, if possible, in an order which is separately randomized for each subject. For illustrative purposes, Figure 10.1 shows the plan of an experiment of this type in which $b = 3$ and $a = 4$. Ignoring the repeated measurements on A, the experiment is simply a random groups design with three levels of B and n subjects in each group. Ignoring the B classification, the experiment is simply a treatments x subjects design with four measurements on each subject, one at each level of A.

Testing the Between-Subjects Factor

For the purpose of evaluating the effect of B, the experiment may be viewed as a random groups design, with the a repeated measurements collapsed into a single score per subject. When the design is looked at in this way, the sum of squares for subjects, with $bn - 1$ *df*,[4] is the equiva-

[4] See Table 10.5 for complete breakdown of *df* for this design.

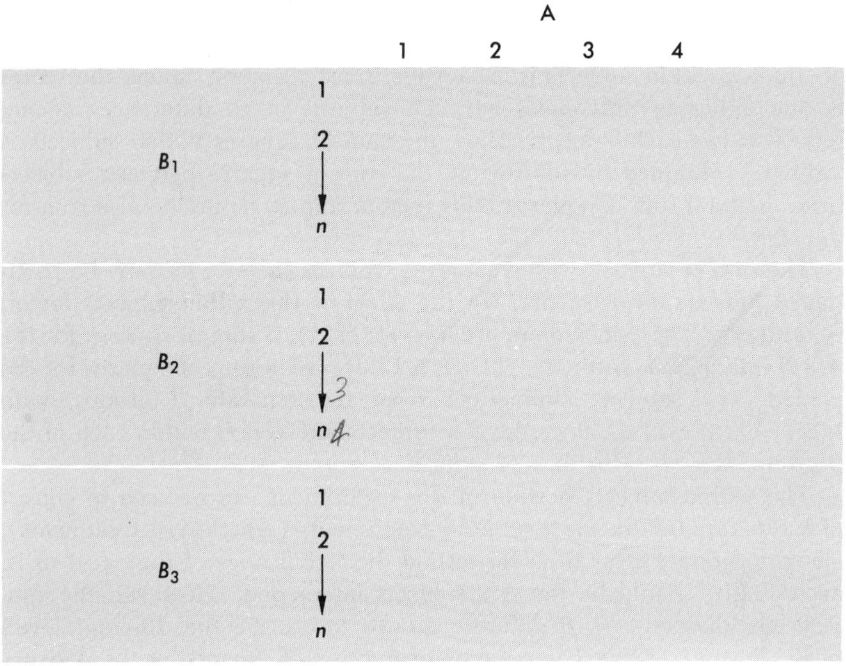

Figure 10.1 Plan of mixed factorial experiment with three levels of B administered between groups and four levels of A administered completely within subjects.

lent of the total sum of squares in a simple random groups design. These between-subjects df may be partitioned into $b - 1$ df for the effect of B (since there are b levels of B) and $b\ (n - 1)$ df for error (since each of the groups has $n - 1$ df between subjects and there are b groups). Also, as is true in any random groups design, the appropriate error term for testing the effect of the between-groups factor, B, consists of the pooled within-groups variance (i.e., the variability of subjects who have been treated similarly). This error term is based, of course, on the assumption of homogeneity of the separate within-groups variance estimates. It should be clear that, aside from the collapsing of the A variable, nothing is done differently in this situation from what would be done in any random groups experiment.

Testing the Within-Subjects Factors

Each subject is measured a times, once at each level of A. Thus, the variability of the a measures has $a - 1$ df within each subject. Since there are bn subjects in the experiment (i.e., n in each of b groups), the pool of all of the within-subjects *df* equals $bn\ (a - 1)$. Partitioning of the total sum of squares into a sum of squares between subjects and a sum of squares within subjects is exhaustive, since variation among the scores is due either to differences between subjects or to differences among scores within each subject. Thus, the sum of squares within subjects is ordinarily obtained by subtracting the sum of squares between subjects from the total sum of squares (this relationship is, naturally, also true for the *df*).

The sum of squares within subjects, with $bn\ (a - 1)$ *df*, may be partitioned into a sum of squares for the effect of the within-subjects factor, *A*, with $a - 1$ *df* (since there are *a* levels of *A*), a sum of squares for the *A* x *B* interaction, with $(a - 1)\ (b - 1)$ *df*, and a sum of squares for the pooled *A* x subjects interactions from the separate *B* groups, with $b(a - 1)(n - 1)$ *df*, since the *A* x subjects interaction within each of the *b* groups has $(a - 1)\ (n - 1)$ *df*.

The within-subjects portion of this experiment can be seen to consist of *b* separate treatments x subjects experiments (where *A* = treatments). The appropriate error term for testing the significance of the effect of *A*, accordingly, should be the *A* x subjects interaction. However, the subjects are divided into *b* different groups to receive the different levels of *B*. Because of this, computation of the sum of squares for the *A* x subjects interaction as if there were no *B* grouping would result in a quantity which is inflated by the possibility that the effect of *A* may differ from one *B* group to another.

This simply means that the overall *A* x subjects interaction contains the *A* x *B* interaction. It is necessary, thus, to obtain the separate *A* x subjects components of the within-error term from within each of the *b* groups and, then, to pool them. This may be done by obtaining the overall *A* x subjects sum of squares (i.e., ignoring the *B* grouping) and then subtracting the sum of squares for the *A* x *B* interaction.[5] Table 10.5 presents the breakdown of *df* for this analysis of variance. The arrows in the table indicate the manner in which the appropriate *F* ratios are formed, as previously.

Because this design uses within-subjects blocking to control for subject heterogeneity in the tests of the within-subjects effects but not in the test of the between-subjects effect, the within-subjects error term is almost

[5] In this design the easiest way to determine the sum of squares for *A* x subjects/within *B* is as a residual, simply by subtracting the sums of squares for *A* and for *A* x *B* from the total sum of squares within subjects. However, in some of the more complex applications of mixed factorial designs presented below the within-subjects error terms cannot be obtained by subtraction but must be obtained according to the principles given here. In the present case, *A* x subjects overall may be obtained as a residual by subtracting $SS_{\text{subjects}} + SS_A$ from SS_{total}.

TABLE 10.5

Breakdown of *df* for analysis of variance of mixed factorial design with one factor between and one factor within subjects.

Source	*df*
Subjects	$[bn - 1]$
→ B	$b - 1$
Error (Ss/within B)	$b(n - 1)$
Within subjects	$[bn(a - 1)]$
→ A	$a - 1$
→ A x B	$(a - 1)(b - 1)$
Error (A x Ss/within B)	$b(a - 1)(n - 1)$

always considerably smaller than the between-subjects error term. Obviously, random differences among different individuals are likely to be greater than random differences within the same individual. Thus, it may often occur that the effect of B may fail to achieve statistical significance while the A x B interaction does, since these tests involve different error terms.

Two Tests of the Same General Hypothesis

When the A factor is sequential (e.g., trials), a significant A x B interaction, even when the main effect of B was itself not significant, is often interpreted to mean that the B factor had a differential effect on the different groups which resulted in different patterns of trial-by-trial responding. There is nothing inherently illegitimate in this procedure, since the test of B and the test of the A x B interaction are statistically independent.

However, if the same general experimental hypothesis is evaluated in these two different ways, an underestimate of the probability of a Type 1 error regarding this hypothesis may result. For example, if the B factor is alleged to have an effect on learning, the experimenter may seek support for this conclusion in a significant main effect of B, which reflects the fact that the different groups tended to respond overall at different levels. In this test, he is concentrating on the *level* of performance achieved in the different groups. He may also seek support for his conclusion regarding learning in a significant A x B interaction, which reflects the fact that the B groups changed differently across trials. In the latter test, the experimenter is concentrating on the *rate* of performance-change in the different groups. Although somewhat different operationally defined manifestations of "learning" are involved in the two different tests, the experimenter may wish to use *either* in support of his general hypothesis regarding learning.

When an experimenter employs the more sensitive $A \times B$ interaction approach to gain a "second chance" to support the same general hypothesis (and when the main effect of B is not significant) it would be appropriate for him to confine himself to statements which are as broad and vague as his procedure of depending upon *either* of two rather different behavioral effects to support the very same behavioral conclusion. On the other hand, if he treats the two different approaches as tests of different experimental hypotheses, one having to do with the overall level of responding and the other with differences in patterns of trial-by-trial responding, neither test is influenced by the other. It is appropriate to reiterate the statement made earlier regarding the experimenter's responsibility to report conclusions and probability levels that are as close to truth as his experimental procedures permit.

Trend Analysis in the Mixed Factorial Design

When the within-subjects variable, A, is administered in equal quantitative steps, the sums of squares for A and for the $A \times B$ interaction may be partitioned into orthogonal trend components in the manner previously discussed in this chapter. The partitioning of the sum of squares for A into its linear, quadratic, etc. trend components is done in the same way as previously described for the treatments x subjects experiment, except that the totals employed are obtained from all of the B groups combined (i.e., the "T" in $(\Sigma kT)^2/n\Sigma k^2$ ignores the B grouping) and $n\Sigma k^2$ becomes $bn\Sigma k^2$, since the mixed factorial experiment contains bn subjects. Partitioning of the $A \times B$ interaction into its trend components is somewhat similar to the procedure for partitioning the treatments x subjects interaction, which was shown above.

The linear component of the $A \times B$ interaction reflects the possibility that the B groups differ across the several levels of A in a linear fashion (i.e., the best fitting straight lines for the various B groups may have different slopes). Similarly, the quadratic component of the $A \times B$ interaction reflects the possible differences in "bowedness" among the different groups' trends (e.g., in a generalization experiment some groups may show steeper gradients than others).

Computation of the sums of squares for the components of the $A \times B$ interaction is achieved by means of the following formula, which will be recognized to be the same as the formula for obtaining the components of the treatments x subjects in interaction, given on p. 217, except for the addition of b, the number of levels of the between groups factor, B:

$$SS_{A \times B \text{ (linear)}} = \frac{\sum_1^b (\sum_1^n \sum_1^a kX)^2}{n\Sigma k^2} - \frac{(\sum_1^b \sum_1^n \sum_1^a kX)^2}{bn\Sigma k^2}$$

Each of the components of the $A \times B$ trend has $b - 1$ df and their sum equals the $(a - 1)(b - 1)$ df of the $A \times B$ interaction.

The appropriate error terms for testing the separate components of the A and the $A \times B$ trends are obtained by partitioning the sum of squares for A x subjects/within B into its orthogonal trend components, $A \times Ss_{lin}$/within B, $A \times Ss_{quad}$/within B, etc. This may be done by computing the $A \times Ss_{lin}$ overall (ignoring the B grouping) and subtracting the $A \times B_{lin}$ from it, by computing the $A \times Ss_{quad}$ overall (ignoring the B grouping) and subtracting the $A \times B_{quad}$ from it, etc. For example:

$$SS_{A \times Ss_{lin}/\text{within } B} = \frac{\sum_1^n \sum_1^b (\sum_1^a kX)^2}{\Sigma k^2} - \frac{(\sum_1^b \sum_1^n \sum_1^a kX)^2}{bn\Sigma k^2} - SS_{A \times B_{lin}}.$$

An alternative and possibly simpler method for obtaining the sum of squares for $A \times Ss_{lin}$/within B, etc., would be to compute its b components separately within each B group and then add them together. Within each B group:

$$SS_{A \times Ss \text{ (linear)}} = \frac{\sum_1^n (\sum_1^a kX)^2}{\Sigma k^2} - \frac{(\sum_1^n \sum_1^a kX)^2}{n\Sigma k^2}$$

Table 10.6 shows the breakdown of df for a trend analysis in a mixed factorial experiment involving one factor between and one equally spaced factor within subjects and indicates with arrows the manner in which appropriate F ratios are formed. The table shows only the df and appropriate tests for the linear and quadratic trend components.

Mixed Factorial Designs Involving Three Independent Variables

Two fairly common extensions of the prototypical mixed factorial design discussed above occur when there are *two* independent variables administered between subjects along with an additional one completely within subjects or when there are *two* completely within-subjects independent variables along with one that is administered between subjects. Although neither of these elaborations of the basic mixed design involves any completely new principles, it will be useful to consider them briefly

TABLE 10.6

Breakdown of *df* for trend analysis in mixed factorial experiment with one factor between and one within subjects, showing only the tests of the linear and quadratic components of trend within subjects.

Source	*df*
Subjects	$[bn - 1]$
→ B	$b - 1$
Error (Ss/within B)	$b(n - 1)$
Within subjects	$[bn(a - 1)]$
→ A_{linear}	1
→ $A \times B_{\text{linear}}$	$b - 1$
$\text{Error}_{\text{linear}}$ ($A \times \text{Ss}_{\text{lin}}$/within B)	$b(n - 1)$
→ $A_{\text{quadratic}}$	1
→ $A \times B_{\text{quadratic}}$	$b - 1$
$\text{Error}_{\text{quadratic}}$ ($A \times \text{Ss}_{\text{quad}}$/within B)	$b(n - 1)$

in turn prior to going on to more complex combinations of blocking between and within subjects.

Of the two extensions in question, the one involving two variables between subjects and one within subjects is by far simpler. Whenever a random groups factorial experiment (with two factors) requires repeated measurements on several occasions or trials, for example, the results may be analyzed as a three-factor mixed design, with trials as the within-subject factor. The between-subjects portion of the experiment is dealt with as if each subject's data were collapsed over the several trials into a single score. As such, it is merely a random groups factorial design.

The within-subjects portion of the experiment contains the main effect of the within-subjects factor, its interactions with each of the between-subjects factors and with their interaction, and the within-subjects error term. The last mentioned component of the within-subjects variability consists of the pooled treatments x subjects interactions from within each of the separate combinations of conditions (i.e., each subgroup) of the factorial design between subjects. Although it is normally obtained in this case as a residual, by subtracting all of the within-subjects effects from the total within-subjects sum of squares, it can also be obtained by computing the treatments x subjects interaction overall, ignoring the between-subjects grouping, and then subtracting the $A \times B$, the $A \times C$, and the $A \times B \times C$ interactions. Table 10.7 summarizes the breakdown of *df*

TABLE 10.7

Breakdown of *df* for analysis of variance of mixed factorial design with two variables between and one within subjects (b = levels of B and c = levels of C).

Source	*df*
Subjects	$bcn - 1$
→ B	$b - 1$
→ C	$c - 1$
→ $B \times C$	$(b - 1)(c - 1)$
Error (subjects/within subgroups)	$bc(n - 1)$
Within subjects	$bcn(a - 1)$
→ A	$a - 1$
→ $A \times B$	$(a - 1)(b - 1)$
→ $A \times C$	$(a - 1)(c - 1)$
→ $A \times B \times C$	$(a - 1)(b - 1)(c - 1)$
Error (A x Ss/within subgroups)	$bc(a - 1)(n - 1)$

for this design, using n for the number of subjects within each of the factorial treatment combinations (each subgroup).

When all of the levels of two of the independent variables are administered completely within subjects and the levels of another variable are administered to different groups of subjects, the resulting mixed design may be viewed as a combination of a within-subjects factorial design and a random groups design. It is as if the within-subjects factorial design were conducted separately in each of the random groups. The between-subjects portion of the experiment is dealt with as is any conventional random groups design, with each subject's repeated measures collapsed over the levels of both the within-subjects factors to produce a single score per subject. The between-subjects error term is the pooled within-groups variance, as in all such situations.

The within-subjects sum of squares (which, as previously, can be obtained as a residual by subtracting the sum of squares for subjects from the total sum of squares) contains a sum of squares for the effect of A, one of the within-subjects factors, another for the effect of B, the other within-subjects factor, another for the $A \times B$ interaction, and three others for the interactions of $A \times C$, $B \times C$, and $A \times B \times C$.

As in the factorial design within subjects, there are three within-subjects error terms. These are A x subjects/within C, B x subjects/within

C, and *A* x *B* x subjects/within *C*. Because there are three error terms, subtraction cannot be used to determine their values. *A* x subjects/within *C* is obtained by computing the *A* x subjects interaction overall (i.e., ignoring the *C* grouping) and subtracting the *A* x *C* interaction from it. *B* x subjects/within *C* is obtained by computing the *B* x subjects overall (ignoring *C*) and subtracting the *B* x *C* interaction from it. The remaining within-subjects error term, *A* x *B* x subjects/within *C*, may be obtained as a residual, by subtracting the sum of all of the previously determined within-subjects sums of squares from the total within-subjects sum of squares. Table 10.8 presents the breakdown of *df* for the mixed

TABLE 10.8

Breakdown of *df* for analysis of variance of mixed factorial design with one factor between and two factors within subjects.

Source	*df*
Subjects	$[cn - 1]$
→ *C*	$c - 1$
Error (Ss/within *C*)	$c(n - 1)$
Within subjects	$[cn(ab - 1)]$
→ *A*	$a - 1$
→ *A* x *C*	$(a - 1)(c - 1)$
Error (*A* x Ss/within *C*)	$c(a - 1)(n - 1)$
→ *B*	$b - 1$
→ *B* x *C*	$(b - 1)(c - 1)$
Error (*B* x Ss/within *C*)	$c(b - 1)(n - 1)$
→ *A* x *B*	$(a - 1)(b - 1)$
→ *A* x *B* x *C*	$(a - 1)(b - 1)(c - 1)$
Error (*A* x *B* x Ss/within *C*)	$c(a - 1)(b - 1)(n - 1)$

factorial design with a single factor between and two factors within subjects, again using arrows to indicate the manner in which appropriate *F* ratios are formed.

Blocking Within Subjects Mixed with Blocking by Grouping

The situation may arise in which individual differences may be controlled in a factorial design by using each subject as his own control with respect to one of the independent variables, while blocks of homogeneous subjects may be employed to control individual differences with respect to a second independent variable. Such a design would be a kind of combination of the prototypical mixed factorial design (with one factor between and one within subjects) and the randomized blocks design. Table 10.9 presents the breakdown of *df* for the analysis of this design. In this table, n refers to the number of blocks and b = the number of levels of B.

TABLE 10.9

Breakdown of *df* for mixed factorial design with one variable within blocks and one variable within subjects.

Source	df *
Subjects	$[[bn - 1]]$
Blocks	$[n - 1]$
B	$b - 1$
Error (B x blocks)	$(b - 1)(n - 1)$
Within subjects	$[[bn(a - 1)]]$
A	$a - 1$
A x B	$(a - 1)(b - 1)$
Error (A x Ss/within B)	$b(a - 1)(n - 1)$

* The double brackets are used to clarify the fact that the sum of squares (and *df*) for B and B x blocks add to the sum of squares for blocks, and that the sums of squares (and *df*) for subjects and within subjects add to the total.

The design shown in Table 10.9 would appear to be useful in experiments in which a group of treatments defining an independent variable were to be compared within blocks of homogeneous subjects, because of the impossibility of administering more than one level of the independent variable per subject (e.g., when the treatments are different training procedures), while at the same time repeated measurement over trials (or some other factor) could be achieved for all subjects. Training procedures, in this example, is, in effect, run within blocks (but between subjects), while trials is run within subjects. As in any randomized blocks design, the error term appropriate for testing the effect of training procedures would be the training procedures x blocks variance estimate, while the test of trials and of trials x training procedures would employ the within-subjects error term, trials x subjects/within training procedures (which, in this case, is also trials x blocks/within training procedures).

It is possible to add another factor to this situation by dividing the blocks into groups of blocks to receive the different levels of a *completely* between subjects independent variable, without adding inordinate complexity to the situation. The experiment would then represent a three-factor design with three different sources of error variance available for the tests of the three different types of effects. One of these error estimates arises from variability between subjects, between blocks; the second from variability between subjects, within blocks; and the third from variability within subjects. The analysis of variance has the superficial appearance of a prototypical mixed factorial design, with one factor between and one within subjects, but with the between-subjects portion subdivided into between-blocks and within-blocks segments. The breakdown of *df* for this design is shown in Table 10.10, with n = the number of blocks within each level of C in this case.

TABLE 10.10

Breakdown of *df* for mixed factorial design with one variable between blocks, one variable within blocks, and one variable within subjects.

Source	*df* *
Subjects	$[[bcn - 1]]$
Blocks	$[cn - 1]$
C	$c - 1$
Error—b_1 (blocks/wC)	$c(n - 1)$
Ss/w blocks	$[cn(b - 1)]$
B	$b - 1$
B x C	$(b - 1)(c - 1)$
Error—b_2 (B x blocks/wC)	$c(b - 1)(n - 1)$
Within subjects	$[[bcn(a - 1)]]$
A	$a - 1$
A x B	$(a - 1)(b - 1)$
A x C	$(a - 1)(c - 1)$
A x B x C	$(a - 1)(b - 1)(c - 1)$
Error—w (A x Ss/wBC)	$bc(a - 1)(n - 1)$

* n = no. of blocks within each level of C.

Experimental Example

An example of the type of experimental situation in which blocking by groups is employed to control individual differences regarding the test of some of the effects, while within-subject blocking is used to control individual differences in the tests of other effects, is found in a study by Bunnell and Kimmel (1965). Male hamsters were matched on preexperimental sexual activity and placed into groups to receive either high or low amounts of sexual experience. These groups were then subdivided such that half of the subjects in each were gonadectomized and half received sham operations. This aspect of the experiment involved a factorial design run within randomized blocks, the blocks being constituted on the basis of level of pre-experimental sexual activity. The two variables run factorially within blocks (between subjects) were amount of sexual experience and whether or not an operation was done. Several postoperative measurement trials were given to each subject, so that within-subject blocking was employed with respect to the effects of trials and all interactions with trials.

Since almost all of the important significant effects in this experiment were found in the within-subjects portion of the analysis, the potential usefulness of experiments employing two different methods of blocking would appear to be considerable. Also illustrated in this example is the fact that the range of elaboration of different types of mixed factorial designs, especially when more than a single basis of blocking exists, is dependent only upon the innovative ability of individual researchers. Once the basic principles of between- and within-subjects blocking are fully grasped, the actual details of each special application may be worked out in a relatively routine fashion.

SUMMARY

One of the most common and effective applications of the blocking principle in psychological experimentation is the use of each subject as his own control in a completely within-subjects design. Both single-factor and factorial experiments can be run in this way, generating what is known as the *treatments-by-subjects* design. As in other instances of blocking, the *treatments-by-subjects interaction* is the *appropriate error term* for evaluating the independent variable's effect.

If there is explicit interest in testing the significance of the interaction, it is necessary to measure each subject at least twice in each condition to provide the appropriate replication for error. In factorial experiments run completely within subjects, the error term is the treatment-by-subjects interaction appropriate for each particular effect in question (i.e., A x Ss, B x Ss, or A x B x Ss). *Trend analysis* in the *treatments-by-subjects* design follows the same pattern as in simpler designs, except the necessary error terms are obtained by partitioning the interaction into *its* linear, quadratic, etc. components.

Mixed factorial designs involve the administration of one or more independent variables to different groups of subjects, as in the random groups design, while one or more others are administered within subjects, as in the treatments-by-subjects design. These designs are probably used more often than any others in psychological research. Tests of the between-subjects effects employ the within-groups variance estimate as an error term, while the within-subjects effects are tested by appropriate treatments-by-subjects interactions (within subgroups). Trend analysis may be employed in these situations also.

Designs involving blocking within subjects mixed with blocking by grouping have also occasionally been employed to advantage. In experiments of this type, three distinctly different sources of error variance may be identified, one completely between subjects regardless of blocking, one between subjects within-blocks, and one within subjects. As in simpler designs, each type of experimental effect is tested with its appropriate error term.

11

Deliberate and Inadvertent Confounding

The term "confounding" has been used frequently in the preceding chapters with only minimal effort towards definition. In Chapter 5, for example, it was suggested that defining an independent variable by selection of different groups of subjects on the basis of a factor such as "economic need" might also accidentally result in the formation of groups that would differ systematically in other extraneous ways that might be relevant to the dependent variable. It was pointed out that the "poor" and "rich" children probably also differed in amount of experience with coins, a difference which could have just as easily influenced the subjects' size judgments as did their "need" for money. In such a case, there would be more than one independent variable which might be systematically influencing the observed behavior, making it impossible for the experimenter to determine *which* factor was actually responsible for the experimental outcome.

The separate influences of these independent variables would be inextricably intertwined, or *confounded.* Whenever manipulation or selection of the levels or categories of an independent variable results in simultaneous systematic variation of unintended additional variables, inadvertent confounding is said to exist. Obviously, no experimenter would choose to permit this type of confounding to occur if he were aware of its potentiality, since his conclusions regarding the effects of the intended independent variable could then not be stated unambiguously. Nevertheless, examples of inadvertent confounding abound in the literature, usually identified in retrospect, pointing again to the difficulty of effective experimental planning.

DELIBERATE CONFOUNDING IN 2^n STUDIES

The dangers of *inadvertent* confounding notwithstanding, it happens that *intentionally* arranged confounding is a device which may be used to achieve reduction in experimental error without an intolerable increase in sample size. Because of its intrinsic value in this regard, and because a thorough understanding of the principles of deliberate confounding may provide a strong safeguard against inadvertent confounding, these principles will be considered in the first part of this chapter. The second part of the chapter is concerned with methods for avoiding or reducing the probability of inadvertent confounding and for dealing with unavoidable but still undesirable confounding.

In introducing the topic of the blocking principle in Chapter 9, brief mention was made of the fact that confounding exists whenever the size of the block is smaller than the size of a replication. Since blocks of subjects or trials define a potentially relevant variable, running some of the treatment-combinations within one block, but others within a different block, would confound the block difference with whatever contrast was defined by the difference between the two sets of treatment-combinations. Thus, the block must be large enough to contain all of the treatment-combinations comprising a replication, in order to avoid inadvertent confounding. Frequently, however, it is impossible to achieve control of a relevant factor in this way, either because the block cannot be made as large as a replication or because it would be unnecessarily wasteful to increase the size of the experiment to achieve this.

To exemplify one such situation, consider a factorial experiment of the 2^n type in which four independent variables (A, B, C, and D) are each administered at two levels, with n subjects randomly assigned to each of the sixteen treatment groups. Suppose that the experiment requires several consecutive days of data collection for each subject and, because of the time required to run each subject, some of the subjects must be run in the daylight hours and some at night.

If substantial differences in the observed behavior may be expected to result from daytime versus nighttime running, randomization of the time of day at which each subject is run on each day would seriously inflate the error variance. Yet, running some of the subjects always during daylight and others always at night would introduce a systematic difference between conditions. It should be noted that the set of subjects run at night constitutes one homogeneous block and the set of subjects run during daylight constitutes another, but both sets are smaller than a complete replication.

Since it is likely that one or more of the fifteen separate single-*df* effects in the four-factor experiment may be of no special interest, and assuming that the diurnal variable (daytime vs. nighttime) is also not an object of study, deliberate confounding may be employed in this situation to great advantage. As was shown in Chapter 8, each of the effects in this 2^n factorial experiment (i.e., *A*, *B*, *C*, *D*, *A* x *B*, *A* x *C*, etc.) consists of a contrast of the data of half of the experimental conditions with the data of the other half.

Procedure for Complete Confounding

Suppose that it has been decided in advance that the quadruple interaction in this study is of no interpretable psychological meaning and, accordingly, useless. The sixteen experimental treatment-combinations may then be divided into two blocks of eight according to the defining contrast for the quadruple interaction, by expanding the polynomial $(a - 1)(b - 1)(c - 1)(d - 1)$ and collecting the positive and negative terms into separate sets, as is shown in Table 11.1.

TABLE 11.1

Defining contrast for *A* x *B* x *C* x *D* interaction.

A x *B* x *C* x *D* +	*A* x *B* x *C* x *D* −
abcd	*abc*
ab	*abd*
ac	*acd*
ad	*bcd*
bc	*a*
bd	*b*
cd	*c*
(1)	*d*

One of the two sets of eight treatments is run in the daytime and the other set is run at night. This assignment is made randomly. In this way, the effect of the diurnal variable is completely embedded, or confounded, into the contrast which defines the quadruple interaction. It will be recalled that each single-*df* effect is orthogonal to each of the others, so that the remaining fourteen effects will be balanced with regard to the diurnal factor. More importantly, experimental error is now not inflated by diurnal variations in behavior, since these are confined completely to the single *df* of the quadruple interaction. The only price paid by the experimenter for this increase in sensitivity in the tests of the various effects of interest is a total sacrifice of information regarding the effect of the quadruple interaction, an effect which was judged in advance to be of no useful meaning anyway. This type of deliberate confounding is known as *complete* confounding, since information regarding the confounded interaction is entirely and inextricably lost.

It is also important to recognize that the increase in sensitivity achieved by controlling the diurnal variable by means of deliberate confounding is far more economical than the alternative of introducing the daytime–nighttime factor as a fifth orthogonal variable in the overall design. Even if the control variable in question is capable of being run as an additional independent factor, to do so would require that the size of the experiment be doubled, or the number of replications within condition halved. Since the diurnal variable is, in the present example, of no interest itself or in its possible interactions with the other four variables, the method of deliberate confounding is obviously preferable.

Simultaneous Double Confounding

In large multifactorial experiments of the foregoing type it may sometimes be desirable to achieve control of *two* potentially relevant variables simultaneously by deliberate confounding. Suppose, for example, that two different experimenters were to be used for data-collection but that neither was regularly available solely for daytime or nighttime work (if they were it would be simple to confound their influence with the already confounded influences of the quadruple interaction and the diurnal variable). If another effect besides the quadruple interaction could be sacrificed, *double* complete confounding could be employed to handle both extraneous control variables, so that neither would inflate the error estimate.

Procedure for Double Confounding

Figure 11.1 illustrates the manner in which this could be achieved, assuming that the $A \times B$ and the $A \times B \times C \times D$ interactions were chosen to be sacrificed.

In Figure 11.1 it can be seen that the eight treatment-combinations in the column labeled *day* compared with the eight in the column labeled *night* give the defining contrast for the $A \times B \times C \times D$ interaction, which is thus completely confounded with the diurnal factor. Similarly, the eight treatment-combinations in the row labeled Exp. 1 and the eight in the row labeled Exp. 2 give the defining contrast for the $A \times B$ interaction, which is thus completely confounded with the experimenter dif-

		A x B x C x D	
		(+) Day	(−) Night
A x B	(+) Exp. 1	abcd ab cd (1)	abc abd c d
	(−) Exp. 2	ac ad bc bd	a b acd bcd

Figure 11.1 Double complete confounding of diurnal and experimenter effects with $A \times B \times C \times D$ and $A \times B$ interactions, respectively.

ference. In terms of actual experimental procedure, what the figure indicates is that experimenter 1 would run a block of subjects in conditions *abcd*, *ab*, *cd*, and (1), in the daytime, and, in addition, he would run a block of subjects in conditions *abc*, *abd*, *c*, and *d*, at night. Experimenter 2 would run a block of subjects in conditions *ac*, *ad*, *bc*, and *bd*, in the daytime, and another block in conditions *a*, *b*, *acd*, and *bcd*, at night.

Generalized Interaction

As can be seen by inspecting the diagonal contrast in Figure 11.1, the interaction between the diurnal and experimenter variables defines still another effect, in this case the C x D interaction. Thus, the C x D interaction is also deliberately confounded by this procedure, whether desired or not. Whenever two variables are simultaneously deliberately confounded with two orthogonal experimental effects, the *generalized* interaction of these effects is automatically also confounded by the interaction of the two other confounded extraneous factors. Identification of the location of the generalized interaction may be achieved by multiplying the letters naming the two confounded interactions and eliminating all squared terms from the product (e.g., $ABCD \text{ x } AB = A^2B^2\,CD$). Even though three of the fifteen single-*df* effects in the four-factor experiment are completely sacrificed in this way, it may often be true that at least three effects are either of no importance or have no psychological meaning.

Partial Confounding

When it is desirable to achieve error-reduction by deliberate confounding of the effect of some extraneous factor, but the experimenter cannot afford to sacrifice information regarding any effect *completely,* it is possible to confound the extraneous factor with one of the higher-order interactions in one replication of the experiment, then to confound it with a different interaction in another replication, etc. This device, which is known as *partial* confounding, permits recovery of information regarding each of the partially confounded interactions from the data in those replications in which it is unconfounded. At the same time, error-reduction is still achieved because the variation due to the confounded variable is confined to the various interactions involved and is still prevented from inflating the error estimate.

Procedure for Partial Confounding

For example, suppose that two different experimenters were involved in collecting the data in the four-factor experiment discussed above. The sixteen treatment-combinations could be divided into two blocks of eight on the basis of the defining contrast for the quadruple interaction, as was done above, so that each experimenter could run a set of eight treatments. This would be done in the first replication of the experiment. For the second replication, the sixteen treatment-combinations would again be divided into blocks of eight, this time on the basis of the defining contrast for one of the triple interactions. Again each experimenter would run one set of eight treatments. For the third replication, another of the triple interactions would be confounded with experimenters, and in the fourth replication, the remaining triple interaction would be confounded with experimenters.

Information regarding each of the four confounded interactions would be available in the data from the three replications in which it was unconfounded, i.e., its sum of squares would be computed using only the appropriate three-fourths of the data. Because the error estimate is obtained from all of the replications, each of which involved confounding of one of the interactions, a special formula for its approximate determination must be employed.[1]

[1] $SS_{\text{error}} = SS_{\text{total}} - 1/N\ (SS_{\text{between treatments regardless of confounding}}) + C'$.

CONFOUNDING THE *A* X *B* INTERACTION IN A 3 X 3 EXPERIMENT

The ingenuity of the experimenter in arranging to reduce error by means of deliberate confounding is not restricted to the 2^n family of factorial experiments. Basically, what is required in all instances of deliberate confounding is that the *df* for the to-be-confounded effect be identifiable and be either directly or indirectly available for deliberate intermixture with the levels or categories of the extraneous variable. As has been shown, direct access to sets of treatment-combinations may be accomplished easily in the 2^n series. In the 3 x 3 factorial, on the other hand, the 4 *df* for the *A* x *B* interaction are not directly available. To get at these *df*, a somewhat more complex procedure must be used. The first step of this procedure involves writing the letters of a 3 x 3 Greco-Latin square into the cells of the factorial *A* x *B* plan, as is shown in Figure 11.2.

		B 1	B 2	B 3
A	1	A_1B_1 $A\alpha$	A_1B_2 $B\beta$	A_1B_3 $C\gamma$
	2	A_2B_1 $B\gamma$	A_2B_2 $C\alpha$	A_2B_3 $A\beta$
	3	A_3B_1 $C\beta$	A_3B_2 $A\gamma$	A_3B_3 $B\alpha$

Figure 11.2 Plan of 3 × 3 factorial with Greco-Latin square written into the cells.

The nine treatment-combinations are then subdivided into three blocks of three by segregating the treatments associated with each of the Latin letters, as is illustrated in Figure 11.3.

Then the nine treatment-combinations are again subdivided into three blocks of three in relation to the Greek letters in the square, as is shown in Figure 11.4.

The 2 *df* associated with the tripartite subdivision based upon the Latin letters (Figure 11.3) are orthogonal to the 2 *df* associated with the tripartite subdivision based upon the Greek letters (Figure 11.4), since each letter in each alphabet occurs only once with each letter of the other alphabet. Since the Latin and Greek letters are also both orthogonal to

LATIN LETTERS		
A	B	C
A_1B_1	A_1B_2	A_1B_3
A_2B_3	A_2B_1	A_2B_2
A_3B_2	A_3B_3	A_3B_1

Figure 11.3 Subdivision of treatments into blocks of three associated with each Latin letter.

GREEK LETTERS		
α	β	γ
A_1B_1	A_1B_2	A_1B_3
A_2B_2	A_2B_3	A_2B_1
A_3B_3	A_3B_1	A_3B_2

Figure 11.4 Subdivision of treatments into blocks of three associated with each Greek letter.

the rows and columns of the square (i.e., to A and B), each subdivision accounts separately for 2 of the 4 *df* of the $A \times B$ interaction. In other words, if the experiment were to be conducted on three days by three different experiments, and if differences among days and experimenters were thought to be sources of substantial potential error variance, the following arrangements would be appropriate to confound both days and experimenters with the $A \times B$ interaction.

Conditions A_1B_1, A_2B_3, and A_3B_2 would be assigned randomly to be run on one of the three days, A_1B_2, A_2B_1, and A_3B_3 would be run on another day, and A_1B_3, A_2B_1, and A_3B_2 would be run on the remaining day. Experimenter 1 would run A_1B_1, A_2B_2, and A_3B_3, experimenter 2 would run A_1B_2, A_2B_3, and A_3B_1, while experimenter 3 would run A_1B_3, A_2B_1, and A_3B_2. The effect of days would be confounded with the Latin 2 *df* of the 4 *df* for the $A \times B$ interaction and the effect of experimenters would be confounded with the Greek 2 *df* of the $A \times B$ interaction.

Thus, two different extraneous factors could be buried in the $A \times B$ interaction and kept out of the error estimate in this way. The $A \times B$ interaction, however, would be totally sacrificed. While it would seldom if ever be the case that the $A \times B$ interaction in a 3×3 factorial experiment could be completely sacrificed, the 3 x 3 interaction in more complicated experiments (e.g., 2 x 3 x 3) might be expendable, leaving unconfounded the 2 x 3 and 2 x 3 x 3 interactions.

QUASI-LATIN SQUARES

In a within-subject factorial design, in which each subject receives all of the treatment-combinations, the order in which these several different conditions are administered may be controlled by employing the Latin square principle. This procedure was described in Chapter 9. As was noted there, it is sometimes impossible to take full advantage of the device of running within-subject factorial experiments in Latin squares because this would require too much experimental time per subject, too much effort per subject, etc.

For example, in a hypothetical experiment on signal detection following elicitation of the orienting reflex by an irrelevant stimulus, the irrelevant stimulus could be in the same or in a different sense modality as the detection stimulus (modality), it could represent either an increase or a decrease in ambient intensity (direction), it could be either a small or a large change (amount), and it could occur either immediately before the detection stimulus or slightly earlier in time (proximity) (cf. Yaremko, 1969).

This 2^4 factorial design generates sixteen different conditions in which detection performance is to be measured. For control purposes, in addition, another sixteen detection stimuli are delivered intermittently without any prior irrelevant stimulation. Because of the possibility that either fatigue or habituation of the orienting reflex, or both, may occur during an overly protracted testing session, it is impossible to employ the 16 x 16 Latin square suggested by these considerations. Nevertheless, it is essential that the *order* of administration of conditions within subjects be controlled. The experimenter is willing to completely confound the quadruple interaction of modality, direction, amount, and proximity of irrelevant stimulation, since it is neither of interest nor would it be easily interpretable. Thus, the sixteen treatment-combinations may be segregated into two blocks of eight on the basis of the defining contrast for the quadruple interaction. Two separate 8 x 8 randomized Latin squares are constructed, each involving one of the sets of eight treatment-combinations which define the quadruple interaction.

Two different sets of randomly assigned subjects would then be run in these quasi-Latin squares. The difference between the two squares (also a difference between two different groups of eight subjects) would be completely confounded with the quadruple interaction, the latter being inextricably lost by design. But the tests of all of the other effects would gain from the control of order of delivery by means of the Latin square arrangement. Table 11.2 shows the breakdown of the *df* for analysis of variance of the results of this quasi-Latin square experiment.

TABLE 11.2

Breakdown of *df* for analysis of variance of quasi-Latin square study of signal detection. Sixteen subjects are run in this study.

Source	*df*
Sequences (subjects)	(15)
Squares *	1
Sequences/w. squares	14
Ordinal positions	7
Modality	1
Direction	1
Amount	1
Proximity	1
M x D	1
M x A	1
M x P	1
D x A	1
D x P	1
A x P	1
M x D x A	1
M x D x P	1
M x A x P	1
D x A x P	1
Error	91
Total	127

* Confounded with M x D x A x P interaction.

Deliberate confounding for the purpose of error-reduction has not been employed widely in psychological experimentation, in the main because few psychologists have had an opportunity to learn about it until fairly recently. It would appear that quite a range of applications exists, both in within-subject factorial designs and in between-subject factorials involving several independent variables. It is fairly common to find in the published literature studies whose overall sensitivity suffers from proliferation of what are actually control variables into what pass as factorial independent variables. Very often factors such as the sex of the subjects, the species of the subjects, etc., are introduced with no a priori interest in their main effects but because their potential interaction with other factors is of considerable interest. Sacrifice of the main effects of such factors, by deliberate confounding with some control variable, frequently will preclude unnecessary increases in the size of the study. In this way more variables of genuine interest can be added. Confounding of *main* effects produces designs which are known as *plaid* designs, because of the appearance from above of agricultural arrangements of this type. Half-plaid designs result from the confounding of main effects and interactions simultaneously. Additional examples of experimental plans utilizing deliberate confounding are presented by Cochran and Cox (1957).

As has been noted already, beyond the direct use of deliberate confounding in psychological research for the explicit purpose of error reduction (i.e., the deliberate construction of homogeneous blocks of smaller size than entire replications), full comprehension of the manner in which deliberate confounding results in loss of information regarding complex interactions may have the prophylactic effect of reducing the frequency of occurrence of inadvertent confounding. Although it is quite true that our greatest concern over inadvertent confounding centers on those situations in which one or more extraneous factors are unwittingly permitted to vary along with explicitly identified independent variables, the potential for a good deal of inadvertent confounding may be found in many experiments in which arbitrary "balancing" of irrelevant variables may accidentally contaminate (partly or completely) some of the complex interactions. When one considers carefully the manner in which some treatments are run in the daytime and some at night, and some by one experimenter and some by another, etc., to achieve deliberate embedding of these variables with interactions, it is easy to recognize how the same consequence could occur unintentionally. At least this type of inadvertent confounding should be avoidable.

UNAVOIDABLE CONFOUNDING

Manipulation of some independent variable often cannot be achieved without simultaneous variation of other factors. A good example of this problem is found in experiments on the effects of extirpation of central nervous system tissue on various aspects of behavior, since at the same time that specific areas of neural tissue are ablated the organism suffers the much more general traumatic experience of the surgery. The question may always be raised whether the observed behavioral effects are due to the specific locus of ablation or to the generalized trauma. It is common in experiments of this type to employ control subjects which undergo a "sham" operation, involving an equivalent degree of general trauma without the removal of neural tissue. Upon recovery from the operation, only the group with specific tissue loss should show effects peculiarly dependent upon the specific structures involved. Behavioral consequences of the trauma of the surgery itself would be expected to be displayed by both groups.

In experiments of this type, the independent variable may be conceived of as the difference between specific tissue removal and nonremoval, with the extraneous factor of traumatic surgery versus nonsurgery unconfounded because of the "sham" operated controls. Another control group which is totally unoperated would reveal the effects, if any, of the surgery *qua* surgery. This method of control is, of course, quite similar to the traditional "placebo" control of drug studies, as is noted in Chapter 12. The principal difference would appear to reside in the fact that the trauma of surgery appears to be an obviously more intense experience than the mere reception of either an orally or parenterally administered chemical substance, although the frequent observation of "placebo effects" testifies to the need for controls of this type in almost all experiments involving chemical or surgical intervention of the organism.

Induced Quantitative Confounding

Quantitative manipulation of an independent variable frequently induces quantitative variation of one or more additional factors, particularly when the independent variable involves variations in time. For example, in studies in which the amount of appetitive reward is varied systematically, subjects given larger amounts of food must also be given longer periods of time in which to eat the food. If confinement in the goal-box area is associated with food-consumption, then subjects which are given larger amounts of food may either benefit or suffer from the additional confinement as well. If successive, "massed" trials are employed in such studies, it may also happen that the groups which receive larger rewards, and which take more time to consume them, also may have slightly longer intertrial intervals and longer total experimental times in each session. Such subjects, in addition, must necessarily have *less* time to spend in their home cages each day, sleeping, grooming, or doing whatever.

Control groups could be run in studies of this type, in which amount of food received is held constant but confinement in the goal-box, or duration of the intertrial interval, is varied. If the behavioral effects of duration of confinement were found to be quite different from those of amount of food *plus* confinement (or if they were found to be nonexistent), this particular source of potential confounding would have been effectively eliminated.

Control by Systematic Variation

This method of control has been called "control by systematic variation" by Bitterman (1965). In his research on phylogenetic differences in learning ability, it has sometimes been impossible to be sure that an observed difference between species in their learning curves is actually attributable to a phylogenetic difference in learning ability rather than being due to some confounded sensory, motor, or motivational variable. For example, it would be quite futile to attempt to control the sensory or motor demands which are placed upon the animals by holding them constant, since no basis for establishing equivalence of stimuli or responses between species exists.

It is possible, however, as Bitterman notes, to conduct control experiments in which systematic variation of the suspected confounding variable is undertaken, to determine whether the particular behavior found in one species may appear in the other *under any level* of the confounding variable whatever. If it does not appear, it may be concluded that a species difference in learning ability was responsible. This is, of course, exactly the method Bitterman has used to demonstrate that successive improvement in a reversal-learning task, shown under a wide variety of experimental conditions by the rat, does not occur under any conditions in fish. This method appears to be useful for dealing with unavoidable confounding in situations specifically concerned with the presence or absence of a particular behavior in a particular species or other class of subjects.

Choosing the Lesser Evil

It is usually all but impossible entirely to eliminate arbitrariness in the manner in which control of extraneous factors is implemented. The experimenter spends a substantial portion of his experimental planning time simply *choosing* between one type of confounding and another, when one but not both can be eliminated. For example, Amsel and Roussel (1952) compared second-runway speeds following either reward or nonreward in the first goal-box, in their pioneer study of the frustration effect. When reward is present in the first goal-box, the animal spends several seconds eating it before it is prepared to run in the second runway. For this reason, the experimenters decided to confine the animal for an equivalent number of seconds on nonrewarded trials. They were, thus, varying not only the sheer presence or absence of "expected" reward, as they intended, but the amount of confinement imposed upon an animal which was ready to run down the second runway.

The frustration resulting from the absence of expected reward, in this study, could easily have been enhanced by a similar effect resulting from confinement itself. But, of course, there would be no way to achieve the needed control in an experiment of this type, except by means of an elaborate control study on the separate effect of duration of confinement. There is little doubt that their choice against permitting immediate exit from the first goal-box on nonrewarded trials was the one most experimenters would have made.

Sometimes different experimenters demonstrate total disagreement regarding their arbitrary choices of unavoidable confounding. In studies of the temporal interval between the conditioned and unconditioned stimuli in classical conditioning, for instance, a choice must be made between permitting the duration of the conditioned stimulus to co-vary along with the temporal interval (e.g., Kimmel & Pennypacker, 1963), on the one hand, or holding the duration of the conditioned stimulus constant but permitting variation of either the time between its offset and the onset of the unconditioned stimulus or variation of the extent to which the conditioned stimulus "overlaps" the unconditioned stimulus (e.g., Hartman & Grant, 1962), on the other.

Both sets of experimenters wished to avoid confounding the difference between delayed and trace conditioning with their intended interval factor. But Kimmel and Pennypacker used a different duration of the conditioned stimulus for each interstimulus interval condition while Hartman and Grant used a constant duration of conditioned stimulus but permitted it to "overlap" the unconditioned stimulus in all conditions except their longest interstimulus interval. In this particular case, the arbitrary choice of which type of unavoidable confounding to permit was apparently not significant, even though the two different sets of experimenters made opposite decisions. This conclusion is based upon the fact that the results of the two studies were in substantial agreement regarding the effect of the principal independent variable manipulated (and confounded differently in each study). Each study may happily be construed to have provided controls for the other's unavoidable confounding, although this fortuitous outcome is far from the rule. More often than not, variation in the manner in which controls for confounding are introduced may end up producing rather different outcomes.

Balanced Two-Way Confounding

Sometimes it may be possible for the experimenter to build into a single experimental plan a device similar to that which was exemplified in the comparison of the Kimmel and Pennypacker with the Hartman and Grant experiments, without an extravagant increase in the size of his study. An example of this method is found in a recent unpublished experiment on habituation of the orienting reflex done in the author's laboratory.

The experimenters were primarily interested in studying the effect on habituation of the orienting reaction to an auditory stimulus of the amount and direction of stimulus change from a previous habituation stimulus. In other words, subjects were to receive a series of presentations of an auditory stimulus until the orienting reflex, as indexed by the galvanic skin response, reached an objective criterion of habituation (i.e., two successive presentations with zero response). Following the achievement of this criterion, a second habituation series was run using a tone which was either 10 or 20 decibels different from the first tone, either stronger or weaker.

A factorial arrangement of amount of change (10 or 20 db) and direction of change (increase or decrease) would permit separate evaluations of the effects of each, as well as permitting a test of their interaction. However, in a simple factorial experiment of this type, the subjects in the different conditions would necessarily differ also with respect to *either* the intensity of the original habituation stimulus or the intensity of the second habituation stimulus.

If a decision were made to habituate all the subjects on the same *original* intensity (i.e., to hold that factor constant), they would have to be rehabituated at different intensities. Thus, any effects of the independent variables of principal interest would be confounded with the potentially more powerful effect of the differences in intensity of the second stimuli (e.g., more intense stimuli require greater numbers of presentations to reach the same criterion of habituation, regardless of either amount or direction of change from the first habituation stimulus).

On the other hand, it could be decided to hold constant the intensity of the second habituation stimulus, while varying the intensity of the original stimulus to make possible the different amounts and directions of shift. But, if this were done, the effects of the shift variables (amount and direction) would be confounded with differences resulting from the fact that different numbers of original habituation trials would have been given to the subjects in the several change conditions. This factor could have an important effect on habituation to the second stimulus.

It should be clear from this discussion that it is patently impossible simultaneously to control *both* the intensity of the original habituation stimulus and the intensity of the second habituation stimulus, while at the same time being able to study the effects of amount and direction of change in a factorial arrangement.

The experimenters ran both types of control at the same time, but in different halves of the overall experiment (i.e., to suffer both types of confounding). Subjects were habituated originally on a 1,000 Hz tone of 40, 50, 60, 70, or 80 decibels. Four times as many subjects were run in the 60 decibel condition than in any of the other four. The subjects who had been habituated with the 40, 50, 70, and 80 decibel tones were then shifted for rehabituation to a 60 decibel tone. The subjects who had been habituated with the 60 decibel tone originally were divided randomly into subgroups for rehabituation with either a 40, 50, 70, or 80 decibel tone.

For the first of these two major groups of subjects (the ones shifted to the 60 decibel tone for rehabituation), amount and direction of shift could be studied with the intensity of the second tone held constant at 60 decibels. For the second of these major groups of subjects (the ones originally habituated on the 60 decibel tone), amount and direction of shift could be studied with intensity of the first tone held constant at 60 decibels. Of more importance, examination of the interactions of the difference between the two subexperiments and the single-*df* effects within each separate experiment (i.e., amount, direction, and amount x direction) would permit an evaluation to be made of the effect of the different methods of control employed. Figure 11.5 illustrates the general plan of an experiment of this type. In Figure 11.5 the arrows indicate the nature of the shift from first to second habituation stimuli.

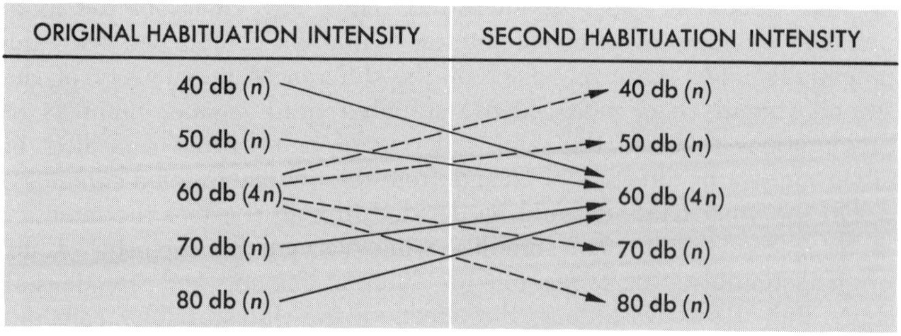

Figure 11.5 Plan of experiment on rehabituation following varying amounts and directions of shift from original habituation intensity.

Table 11.3 shows the mean numbers of trials needed to reach the second habituation criterion in the various groups in this experiment.

As the data in Table 11.3 show, the results of the two subexperiments were quite similar but not completely identical. Both revealed an interaction between direction and amount of shift, since in both subexperiments the intensity factor was positively related to numbers of trials to the habituation criterion when the direction of change was upward, but was negatively related to numbers of trials to criterion when the direc-

TABLE 11.3

Mean numbers of trials to rehabituation following different amounts and directions of shift in intensity, with either original intensity or second intensity held constant.

	Shifted to 60 db		Shifted from 60 db	
	10 db	20 db	10 db	20 db
Increase	8.6	11.3	9.6	20.9
Decrease	11.0	2.6	10.1	7.6

tion of change was downward. The interaction was somewhat more pronounced when the second stimulus was held constant than when the first stimulus was held constant, but this triple interaction (amount x direction x subexperiments) fell far short of statistical significance. In this particular experiment, accordingly, the technique of employing both types of control simultaneously as an additional orthogonal factor in the experimental plan proved to be most useful, since it revealed the fact that the observed interaction between amount and direction of shift was not due to uncontrolled variation *either* in original or in second stimulus intensity. This statement may be made because the interaction in question occurred equally when each intensity was separately held constant in the two subexperiments.

Baseline Data

Two additional features of designs involving change from one condition to another are deserving of comment. Some researchers have stressed the importance of obtaining "baseline" data (e.g., Sidman, 1960) before the introduction of the independent variable under study, basing their argument on the apparent power of this application of J. S. Mill's traditional Method of Differences.

Acceptance of the conclusion that nothing but the independent variable was responsible for the observed change in behavior does seem on the surface to be more straightforward when a steady state existed prior to its introduction. However, it may sometimes be the case that the procedures required for obtaining the original steady state are such as to prepare the subject to be influenced by the independent variable in a particular way, much in the fashion that a pretest may sensitize subjects to be influenced by a subsequent experimental treatment (cf. Chapter 3).

The only thing that can be done to ensure that the baseline condition did not influence the later effect of the independent variable, unfortunately, would be to employ different baseline conditions (e.g., different baseline reinforcement schedules) for different subjects in the experiment. As has been demonstrated recently (Weiner, 1964, 1969), prior schedules of reinforcement definitely influence the effects of current schedules, sometimes to such an extent that it is impossible to comprehend the effects of a current schedule without knowledge of the subject's prior history of reinforcement schedules.

Contrast Effects

In addition to the foregoing type of potential contamination influence, it is also often true that the actual effect of an independent variable may be substantially different when it is studied within subjects rather than between subjects. Grice and Hunter (1964) have recently described "contrast" effects of this type in studies of reaction time and human eyelid conditioning.

When different groups of subjects are conditioned using different intensities of an auditory conditioned stimulus, with each subject receiving only a single intensity, hardly any difference in strength of conditioning is found among the different groups. But, when individual subjects serve as their own controls and receive two or more levels of the independent variable, the stronger conditioned stimuli are significantly better than the weaker ones in eliciting conditioned responses. This effect is not entirely due to the increased sensitivity of the within-subjects design, as Grice and Hunter's data show, since as much of the increased effect appears to be due to an actual increase in the observed difference between the different stimuli as is due to a reduction in the size of the error estimate. In fact, in their reaction time study, Grice and Hunter found that the within-subject experiment resulted in slower reactions to the weaker stimuli as well as faster reactions to the stronger stimuli. It was for this reason that they referred to the effect as a "contrast" phenomenon and suggested that its explanation might be found in the concepts of adaptation-level theory. Certainly, in research involving human subjects, and probably in studies of all higher organisms, within-subject relational effects should not be lightly dismissed by experimenters whose aim is to state general laws applicable to both between- and within-subject situations.

It is, of course, much too easy simply to urge that inadvertent confounding be avoided at all times and that unavoidable confounding also be dealt with intelligently. That these admonishments are not in themselves sufficient to eliminate ambiguous experimental results is demonstrated by the ubiquity of published occurrences of confounding. From the present viewpoint, increased use of deliberate confounding in psychological research may provide an effective prophylactic, since the experimenter who has gained experience in working through the planning stages of studies employing deliberate confounding is likely to be best prepared to avoid accidental accomplishment of similar but contaminating effects.

SUMMARY

Whenever manipulation or selection of the levels of an independent variable results in simultaneous systematic variation of additional relevant variables, *confounding* is said to exist. When this occurs inadvertently, interpretation of the results of an experiment may be difficult or even impossible. However, *deliberate* confounding is a special instance of the blocking principle, in which the size of the block is smaller than the size of the replication, and which may be used as an error-reduction device in experiments with one or more effects of no interest. In such cases, the *defining contrast* of the effect to be confounded is identified and the levels of the extraneous variable are intermixed with it. This has the desired consequence of removing variability due to the extraneous factor from the error term and placing it inextricably in the defining contrast and degrees of freedom of the confounded effect. In this way one or more control variables may be kept from inflating the error variance. When two effects are simultaneously confounded deliberately, their *generalized interaction* is also confounded by the interaction of the two control variables, which may sometimes create a problem.

In certain cases it may be possible to employ *partial confounding*, in which the control variable is intermixed with different effects from replication to replication, so that each partially confounded effect may be evaluated separately in that portion of the data in which it is unconfounded. These procedures are most applicable in studies involving factors with two levels each, but may be used in more complex designs as well.

When the Latin square design is to be used to control order effects in within-subjects factorial experiments, but there are too many treatments to be given to each subject, because of fatigue, boredom, etc., the defining contrast for one of the effects may be used to divide the number of treatments in half for each subject. Then two *quasi-Latin squares* are employed and the extraneous control variable is confounded with the difference between the two squares, which also identifies the confounded effect.

It frequently occurs that the experimenter recognizes that manipulation of an independent variable results in *unavoidable confounding* of one extraneous factor or another, depending upon his choice of how to achieve the manipulation. This may sometimes be dealt with effectively by *systematic variation* of the control factor of concern over most of its range, so that it may be eliminated from consideration as a source of contamination.

It may also be possible to employ special designs in which confounding by one extraneous factor exists in part of the study and confounding by another is present in another part. The effect of interest is then evaluated in both parts of the study together to see whether it changes as a consequence of which extraneous factor is confounded. This problem is most likely to be troublesome when sequential within-subjects manipulations are employed, since these are most likely to involve possible *contrast* effects.

12

Concluding General Principles

CONDUCTING SUCCESSFUL EXPERIMENTS

No simple formula exists for conducting successful experiments. Slogans, such as "Hold everything constant except the variable of interest," or "Establish a stable baseline and then administer the independent variable," are nothing more, in the final analysis, than slogans. No substitute can be found for originality, none for thoughtful planning, and none for flexibility. Complete mastery of all of the design techniques presented in this or any other book will assure no one of success as an experimenter, just as perfect spelling and grammar, and a large vocabulary, are insufficient to produce a successful writer. But, while not a *sufficient* condition, competence in the principles of experimental design is one of the *necessary* conditions which, when added to inventiveness, persistence, and good luck, may result in good experimentation.

Observation, Observation, Observation

Among the more important general principles which may be of assistance in guiding researchers to potentially fruitful experimentation, none deserves more emphasis than the deceptively simple reminder that psychology is, first and last, an *empirical* science. Its questions are scientifically meaningless unless they can be answered by means of concrete *observations*, its theoretical concepts are of value only to the extent that they are anchored firmly in and have implications for actual *observations*, and the results of its experiments, finally, are significant only when they can be replicated by the *observations* of other experimenters. The full implications of this simple reminder are too numerous to be listed. Perhaps its flavor may best be communicated by the dictum, "Never argue about whether or not it's raining. Look out the window!" Or maybe Pavlov's comments on this point are cogent, "Although a bird's wing is perfect, the bird could never soar if it did not lean upon the air. Facts are the air on which the scientist leans." (Kaplan, 1966, p. 385.)

At the risk of seeming pious, it is necessary to note also the importance of *patience* in doing experimental research. Most experiments involve a great deal of tedious effort and many may be quite frustrating. Many of the most important discoveries in science have been the result of long series of arduous experiments, each of which separately is an essential element of the whole endeavor, each of which adds discernibly, albeit minutely, to the developing state of knowledge. Even experiments with negative results are of potential scientific usefulness, if they succeed in invalidating a false hypothesis or if they contribute to the planning of subsequent, more successful studies.

A corollary of this injunction against impatience and haste is the "commandment," "Thou shalt not be greedy!" The temptation to try to answer as many questions as possible in a single experiment has probably been the cause of many avoidable experimental failures. What is referred to here is the tendency to add independent variables, either by subdividing the experimental groups into smaller subgroups or by employing repeated measurements of the same subjects under several different conditions, after the *basic* experimental purpose and plan have been established. Because of the substantial initial investment of experimenter time, laboratory space and equipment, and experimental subjects in the conduct of any sound psychological experiment, experimenters sometimes succumb to greediness and add experimental manipulations beyond those originally planned, in an attempt to obtain as much information as possible in return for their investment. But the addition of new variables may reduce the sensitivity of important comparisons by reducing the size of the subgroups or increasing variability. It may complicate irremediably both the experimental procedure and the interpretation of possible outcomes. And it may change the experimental situation for the subject, when repeated measures are required by the additional procedures, so that fatigue, boredom, or similar undesirable consequences may result. There is a difference between adding experimental comparisons because of careful problem analysis, on the one hand, with the effect of putting the experimenter's conceptualizations into greater empirical jeopardy (a desirable state of affairs), and, on the other hand, simply trying to pick up any additional scraps of *possibly* useful information (which may have many undersirable consequences).

Flexibility

Patiently pursuing a continuously unfolding and developing experimental program, however, should never have the effect of placing blinders over the eyes of the experimenter, with the effect of reducing his sensitivity to unexpected, paradoxical, or otherwise intriguing results. As Skinner (1956, p. 223) has remarked (exaggeratedly), "When you run onto something interesting, drop everything else and study it." The word "interesting" in Skinner's statement, of course, may have different meanings for different scientists. In the present frame of reference, it may be taken to mean "inexplicable in terms of known principles," or, sometimes, "conflicting with known or allegedly known principles." When an experimental observation is either inexplicable or seemingly in conflict with what is presumably already known (especially including the presuppositions and untested assumptions which serve as a foundation for the experimental program), there is nothing more important the experimenter can do than turn his attenion to it. No current theoretical pursuit in which the experimenter may be engaged can take priority over the resolution of inconsistencies related to his *basic* assumptions.

For example, given the fundamental assumption that the *complex* behavior of living organisms is capable of analysis in terms of principles derived from research on its *simpler* elements, how could an experimenter not be perplexed and intrigued by an observation such as Pavlov's (1960) finding that apparently neurotic behavior is displayed by dogs that are confronted with differentially reinforced but objectively highly similar stimuli? In this type of experiment, a visual stimulus, a circle, is presented repeatedly and followed each time by food, while another stimulus, an ellipse, is presented equally often but without food. The dog comes to make an anticipatory salivary reaction when the circle is presented but not when the ellipse is presented. The experimenter now gradually begins to reduce the eccentricity of the ellipse, trial after trial, so that it becomes more and more similar to the circle, still continuing to present the food reinforcement only with the circle. When the ratio of the two axes of the ellipse is reduced, finally, to about 9:8, the dog begins to salivate indiscriminately and copiously to the two stimuli, it whines and trembles, and it attempts to leave the experimental situation. Outside of the experimental situation, it continues to display "neurotic" (i.e., emotional) behavior and cannot be used in experiments again.

The simple breakdown of discriminative responding is hardly surprising. In fact, Pavlov first undertook research of this type because he was interested in employing his conditioning method to measure the discriminative abilities of the dog. It was expected that gradual alteration of the ellipse in the direction of the circle would finally result in a failure of the dog to respond differentially. But, nothing immediately apparent in the training and testing situation prepared the experimenter to expect the dog's behavior to become so generally pathological at the same time that it could no longer respond differentially to the two stimuli. After all, the delivery of the food reinforcement during training did not depend upon the dog's making a "correct" response. Food was simply always *given* after the circle but never after the ellipse.

From an anthropomorphic viewpoint it might be asked, "Why should it make any difference to the dog whether the two stimuli are very different or similar, since there is no practical reason to attend to them anyway?" In other words, the phenomenon did not at first seem to fit into what was then known either about dogs' behavior or conditioning principles. In addition, its appearance seemingly conflicted with the presupposition that experimentally derived principles regarding simple behavioral elements could be applied to more complex behavior. Yet, here was a rather complex pattern of disturbed emotional behavior, apparently resulting from the dog's exposure to the experimental situation, yet not easily predictable from what was done in the situation. To be sure, Pavlov had run onto something interesting and he immediately attempted to study it and comprehend it, although he did not stop everything else he was doing.

Sometimes an experimenter's interest is drawn to an unexpected phenomenon because it conflicts with what *other* experimenters have assumed about the organization of behavior. For example, Maier (1949) found that rats develop rigid and seemingly maladaptive response tendencies in inconsistent reward–punishment situations when they have first learned adaptive responses under consistent reward–punishment conditions. His fixation on this phenomenon was motivated by the belief that he had found a major chink in the presuppositional armor of associative behaviorism. He, thus, pointedly referred to the phenomenon as "behavior without a goal." In this case, it remained for others to demonstrate that principles derived from simple behavioral experiments could account for his complex findings (Wilcoxon, 1952). Other examples of this type, by no means restricted to the field of experimental psychopathology, are Kohler's (1947) observation of transposition in chickens, Olds and Milner's (1954) discovery of a "reward" center in the rat's brain, Skinner's (1956) accidental observation of the partial reinforcement effect, and Gibson's finding of figural aftereffects (Gibson, 1933). Each of these exemplify observed phenomena apparently in conflict with the basic assumptions of many experimental psychologists, at least at the time they were first reported.

Serendipity

Significantly, sensitivity to unexpected or inexplicable findings is likely to be greatest in situations in which a highly systematic approach has been taken, whether or not it is a highly abstract theoretical approach. This relationship is only to be expected, since novelty and unexpectedness presuppose the existence of some kind of organized system of knowledge against the background of which an event can be judged as novel or unexpected. Perhaps not as obviously, unusual findings do not arise because of a careless or sloppy scientific methodology, nor are they as likely to be noticed by casual as by passionate experimenters. There is, in general some misunderstanding about the role of luck in scientific discovery, possibly partly because of the frequency with which experienced but modest experimenters mention it. *Serendipity,* the art of accidentally finding something other than that which was being sought, comes most often to those investigators whose probing activities persist most indefatigably and whose sensitivity to the novel or inexplicable is most keen.

Support for the assertion that serendipitous findings are more likely to occur in an organized experimental context may be found in the observation that methodological devices intended to enhance experimental control often are the catalysts for accidental discovery. It has become part of the folklore of experimental science that the "control group often turns out to be the best experimental group." Methodological serendipity of this type is exemplified by the occasional finding that presumably null or placebo procedures which are administered to control groups may have significant, unexpected effects of their own. The procedure of presenting rewarding events during periods of nonresponding, used as a control in studies of the effects of response-contingent rewards, for example, may result in the acquisition of a learned tendency toward reduced responding, rather than having no learning effects whatever, as expected (Kimmel, 1967). The methodological concern which leads to the use of control procedures of this type results also in the accidental finding that they have effects of potential importance.

SOME ADDITIONAL REMARKS ON CONTROL GROUPS

Basically, control groups are employed to ensure that the specific treatment given to the experimental groups is responsible for the behavior observed in its subjects. As was noted in Chapter 11, experimental treatments involving significant intervention into the ordinary activities of the subjects may result in changes in behavior simply because of the general trauma or disruption occasioned by the intervention. This is particularly likely when the intervention is surgical or chemical, in which case "sham" operated or "placebo" controls are almost always needed.

Even the simple fact that the experimental subject receives special attention or, in the case of subhumans, handling, can produce behavioral changes which may be erroneously attributed to the independent variable. The famous Hawthorne studies (Blum & Naylor, 1968) showed that an *atmosphere* effect could be created by almost any intrusion into the daily working environment of industrial employees, such as bringing in men to paint the walls of the shop or to change the lighting. Increased production, which at first was thought to be the result of the improved environmental conditions resulting either from the painting or from the changed illumination, was found ultimately to be attributable merely to the fact that management had shown concern and attention for the worker's environment rather than to the new color of the walls or the changed lighting. In fact, production was shown to increase *again* following the restoration of the original environmental conditions.

It Takes a Difference To Make a Difference

In devising appropriate procedures for control groups, the experimenter makes an effort to emulate all but the presumably *essential* aspects of the experimental treatment. As was noted in Chapter 10 (p. 208), it may sometimes occur that certain inessential differences between what are initially conceived to be the experimental and control conditions cannot be eliminated without seriously weakening the strength of the independent variable, because these differences turn out to be inseparably correlated with the essential differences.

This problem was present in the experiment on the effect of value on the perceived size of objects, which was discussed in Chapter 5. It will be recalled that an effort was made in Landsman's (1964) study to overcome the difficulty of the earlier Bruner and Goodman (1947) experiment which resulted from the use of cardboard discs as nonvalued control stimuli. The cardboard discs differed from the coins in several ways other than value, for example, in color, in reflectance, in surface texture, etc. As a matter of fact, the only way in which the coins and discs did *not* differ was in size and shape. Landsman used comparison discs made out of coin metal, the alloy used to make actual coins. He engraved horizontal and vertical grid lines on the discs so that the reflected light from their surfaces would be diffused in the same general way light reflected from a coin is diffused by the engraving on its surface. All of this was done to make the noncoins look like coins as much as possible, *without actually being coins and having value.* The danger, obviously, is that the discs may get to look so much like coins that they are responded to as if they were coins. The problem is that there are certain features of coins which are inherent in the definition of what a coin is. The more of these features that are given to the noncoins, the weaker is the independent variable based upon the essential difference between valued coins and nonvalued discs. It takes a difference to make a difference, a point emphasized by the late Professor M. F. Metfessel of the University of Southern California.

Careful analysis of an experimental condition which is assumed to be essential to the definition of an independent variable, so that an effective control procedure may be devised for comparison with it, sometimes results in the realization that it is impossible to verify experimentally that the condition really is essential. For example, it has been proposed that the only essential difference between Pavlovian classical conditioning and Thorndikian instrumental conditioning upon which researchers in this area would agree is the procedural one of whether or not the delivery of the reinforcement is *contingent* upon the performance of the response (Church, 1969). In instrumental conditioning, it has been alleged, the response *produces* the reinforcement (e.g., the rat's depression of the lever in a Skinner box mechanically operates the food-delivery magazine), but in classical conditioning the reinforcement is delivered independent of the subject's response. It has been argued, however, that the essential contingency ("as far as the subject is concerned") is *temporal* rather than *productive.* That is, the reinforcement simply *follows* the response and that is what is essential (Razran, 1965; Skinner, 1953).

In the context of the present discussion, the question is whether a control procedure can be devised for comparison with the conventional instrumental conditioning procedure, in which the reinforcement simply follows the response (i.e., is temporally contingent upon it) but is *not* produced by it. Consider the possibilities. If yoked control subjects, which received reinforcements only when the experimental subjects made responses, were run, the degree of temporal contingency in the two conditions would differ. The experimental subjects would have a greater number of their responses closely followed by reinforcement than would the control subjects. The only way to equate the subjects in degree of temporal contingency is to monitor the control subjects' responses, either mechanically or via the intermediation of the behavior of the experimenter. Compare, for example, the manner in which reinforcements are administered by the experimenter in verbal operant conditioning experiments with the more mechanized method of the Skinner box. If "produces" means "causes," temporal contingency is identical to productive contingency. Thus, no experimental comparison between the two can be made. The question, "Does the reinforcement have its behavior-modifying effect because the response *produces* the reinforcement or, simply, because it is *followed* by the reinforcement closely in time?" turns out to be operationally meaningless, since analysis indicates that there is no way to differentiate these procedurally.

Nevertheless, there is no more empirical justification for the argument that it is merely the temporal contingency that makes the reinforcement effective in strengthening responding than there is that productive contingency is what is necessary. They are simply one and the same operationally. The former possibility happens to seem to be more parsimonious, but that would appear to be a result of the ways in which psychologists have learned to theorize. Human subjects frequently verbalize regarding the productive contingency apparent to them in instrumental conditioning experiments, although these verbal responses may simply be just another part of what the subject acquires in the experimental situation. In studies in which the response-reinforcement relationship is camouflaged by the experimental procedure, in fact, superstitious patterns of responding may be acquired at the same time that the subject acquires the verbalized belief that these patterns bear a productively contingent relationship to the reinforcement (e.g., Timmons, 1963). In experiments of this type, neither the superstitious pattern of responding nor the verbalized belief regarding its necessity actually "produce" the reinforcement. But, "as far as the organism is concerned," the pattern is necessary.

THE CHALLENGE OF EXPERIMENTATION

There is something to be said in favor of a modest degree of immodesty on the part of experimenters. Unless the scientist takes his own hunches seriously, he is not likely to put them to experimental test. Nor will anyone else, if they are not seriously presented publicly. Furthermore, it is only by proceeding on the assumption that his conceptualization is essentially correct that he can engage in productive problem analysis, as has been noted previously. The tendency to be overly modest regarding one's own ideas, it should be noted, also has the effect of protecting them from exposure to empirical scrutiny. But, this is precisely what experimental science is *not* about.

Each new experiment should constitute a real challenge to the researcher. It is better to have tried out a bad idea experimentally and discovered in that way that it was a bad idea than to engage in study of only the completely obvious and predictable. It is reported that C. L. Hull used to make bets with his colleagues and students regarding the outcomes of their experiments (the stakes of these bets are said always to have been one chocolate malt). Whether or not explicit bets are made, there should always be an aspect of experimentation that can be characterized as "living dangerously."

This notion that each experiment should represent something of a challenge to the experimenter is relevant principally to hypothesis-testing studies. As has been observed previously, estimation experiments involve far more predictable events than do those in which hypotheses are tested, in the sense that the former studies presuppose that a relationship between the independent and dependent variables has already been established in prior experiments of the latter type. Nevertheless, even the most precise estimation experiment involves the considerable challenge of establishing the degree of experimental control necessary for accurately determining the shape of a quantitative relationship between variables. To achieve this level of control, the experimenter may have to hold constant a number of relevant variables and deal effectively with individual differences among subjects whether it be by the use of homogeneous blocks of subjects or by within-subject control.

In the final analysis, the business of scientific experimentation must be appreciated in relation to other than technical scientific criteria. In concluding this book with some comments about the challenge of doing experiments, the author is thinking as much of the human experimenter as he is the technical discipline. As Tolman (1959) said, in commenting on his long career as an experimental psychologist, "Since all the sciences, and especially psychology, are still immersed in such tremendous realms of the uncertain and the unknown, the best any individual scientist, especially any psychologist, can do seems to be to follow his own gleam and his own bent, however inadequate they may be. In fact, I suppose that this is actually what we all do. In the end, the only sure criterion is to have fun." (1959, p. 152.)

SUMMARY

It cannot be overemphasized that the success or failure of the science of psychology rests upon the basic empirical observations made in its research laboratories. Sophisticated methods, eleborate theories, etc., in the long run all must take second place in importance to the fundamental data-base. *Facts* are the air on which the scientist leans. Sometimes significant observations occur accidentally, but more often they result from patient and organized experimentation. The curiosity of the scientist and his sensitivity to unusual or inexplicable observations must remain high at all times. Finally, the business of devising and conducting experiments whose goal is to reduce man's uncertainty regarding nature should be both a challenge and a highly satisfying enterprise.

References

AMSEL, A., & ROUSSEL, J. Motivational properties of frustration: I. Effect on a running response of the addition of frustration to the motivational complex. *Journal of Experimental Psychology,* 1952, **43**, 363–368.

BAXTER, B. Problems in planning of psychological experiments. *American Journal of Psychology,* 1941, **54**, 270–280.

BITTERMAN, M. E. Phyletic differences in learning. *American Psychologist,* 1965, **20**, 396–410.

BLUM, M. L., & NAYLOR, J. C. *Industrial psychology.* New York: Harper and Row, 1968.

BONEAU, C. A. The effects of violations of assumption underlying the *t* test. *Psychological Bulletin,* 1960, **57**, 49–65.

BORING, E. G. The role of theory in experimental psychology. *American Journal of Psychology,* 1953, **66**, 169–184.

BOX, G. E. P. The exploration and exploitation of response surfaces—some general considerations and examples. *Biometrics,* 1954, **10**, 16–60.

BOX, G. E. P., & YOULE, P. V. The exploration and exploitation of response surfaces —an example of the link between the fitted surface and the basic mechanism of the system. *Biometrics,* 1955, **11**, 287–323.

BRIDGMAN, P. W. *The logic of modern physics.* New York: Macmillan, 1928.

BROWN, J. S. *The motivation of behavior.* New York: McGraw-Hill, 1961.

BRUNER, J. S., & GOODMAN, C. C. Value and need as organizing factors in perception. *Journal of Abnormal and Social Psychology,* 1947, **42**, 33–44.

BUNNELL, B., & KIMMEL, M. E. Some effects of copulatory experience on postcastration mating behavior in the male hamster. *Psychonomic Science,* 1965, **3**, 179–180.

CHURCH, R. M. Systematic effect of random error in the yoked control design. *Psychological Bulletin,* 1964, **62**, 122–131.

CHURCH, R. M. Classical and instrumental conditioning. In D. R. Jacobs (Ed.), *The behavior therapies.* Cleveland: Case Western Reserve University Press, 1969.

COCHRAN, W. G., & COX, G. M. *Experimental designs.* New York: Wiley, 1957.

COMREY, A. L. A proposed method for absolute ratio scaling. *Psychometrika,* 1950, **15**, 317–325.

COX, D. R. *Planning of experiments.* New York: Wiley, 1958.

CRONBACH, L. J. The two disciplines of scientific psychology. *American Psychologist,* 1957, **12**, 671–684.

DENNY, J. P. Effects of anxiety and intelligence on concept formation. *Journal of Experimental Psychology,* 1966, **72**, 596–609.

DOBRONRAVOV, S. N., & FISHELEV, YA. R. On "dermal sight." *Byulleten Eksperimental'noy Biologii i Meditsiny,* 1964, **8**, 13–16.

DUNCAN, D. B. Multiple range and multiple *F* tests. *Biometrics,* 1955, **11**, 1–42.

DUNNETT, C. W. A multiple comparison method for comparing several treatments with a control. *Journal of the American Statistical Association,* 1955, **50**, 1096–1121.

EBBINGHAUS, H. *On memory.* Leipzig: Duncker & Humblot, 1885.

EDWARDS, A. L. *Experimental design in psychological research.* New York: Rinehart, 1960.

FISHER, R. A. *The design of experiments.* New York: Hafner, 1953.

FISHER, R. A., & YATES, F. *Statistical tables for biological, agricultural, and medical research.* London: Oliver & Boyd, 1953.

GIBSON, J. J. Adaptation, aftereffect, and contrast in the perception of curved lines. *Journal of Experimental Psychology,* 1933, **16**, 1–31.

GORMEZANO, I., MOORE, J. W., & DEAUX, E. Yoked comparisons of classical and avoidance eyelid conditioning under three UCS intensities. *Journal of Experimental Psychology,* 1962, **64**, 551–552.

GRANT, D. A. Analysis-of-variance tests in the analysis and comparison of curves. *Psychological Bulletin,* 1956, **53**, 141–154.

GRANT, D. A. Testing the null hypothesis and the strategy and tactics of investigating theoretical models. *Psychological Review,* 1962, **69**, 54–61.

GRICE, G. R., & HUNTER, J. J. Stimulus intensity effects depend upon the type of experimental design. *Psychological Review,* 1964, **71**, 247–256.

GRICE, G. R., & SALTZ, E. The generalization of an instrumental response to stimuli varying in the size dimension. *Journal of Experimental Psychology,* 1950, **40**, 702–708.

GUILFORD, J. P. *Psychometric methods.* New York: McGraw-Hill, 1954.

GUTTMAN, N., & KALISH, H. I. Discriminability and stimulus generalization. *Journal of Experimental Psychology,* 1956, **51**, 79–88.

HAGGARD, E. A. *Intraclass correlation and the analysis of variance.* New York: Dryden Press, 1958.

HARLOW, H. F. The nature of love. *American Psychologist,* 1958, **13**, 673–685.

HARTMAN, T. F., & GRANT, D. A. Differential eyelid conditioning as a function of the CS-UCS interval. *Journal of Experimental Psychology,* 1962, **64**, 131–136.

HERRNSTEIN, R. J., & HINELINE, P. N. Negative reinforcement as shock-frequency reduction. *Journal of the Experimental Analysis of Behavior,* 1966, **9**, 421–430.

HILL, F. A. Effects of instructions and subject's need for approval on the conditioned galvanic skin response. *Journal of Experimental Psychology,* 1967, **73**, 461–467.

HOVLAND, C. I. The generalization of conditioned responses. I. The sensory generalization of conditioned responses with varying frequencies of tone. *Journal of Experimental Psychology,* 1937, **17**, 125–148.

HULL, C. L., FELSINGER, J. M., GLADSTONE, A. I., & YAMAGUCHI, H. G. A proposed quantification of habit strength. *Psychological Review,* 1947, **54**, 237–254.

JACOBS, A. E. Formation of new associations to words selected on the basis of reaction time-GSR combinations. *Journal of Abnormal and Social Psychology,* 1955, **51**, 371–377.

JENKINS, J. G., & DALLENBACH, K. M. Oblivescence during sleep and waking. *American Journal of Psychology,* 1924, **35**, 605–612.

KAPLAN, M. (Ed.) *The essential works of Pavlov.* New York: Bantam Books, 1966.

KIMMEL, E. Judgments of UCS intensity and diminution of the UCR in classical GSR conditioning. *Journal of Experimental Psychology,* 1967, **73**, 532–543.

KIMMEL, H. D. Instrumental conditioning of autonomically mediated behavior. *Psychological Bulletin,* 1967, **67**, 336–345.

KIMMEL, H. D., BOICE, C., & LECKART, B. T. Paired comparison scaling in two different experiments on habituation. *Behavioral Research Methods and Instrumentation.* 1969, **1**, 107–109. (a)

KIMMEL, H. D., BOICE, C., & LECKART, B. T. Paired comparison scaling of reaction potential from simultaneous measurements of GSR and looking time. *Perception and Psychophysics,* 1969, **5**, 294–296. (b)

KIMMEL, H. D., & KIMMEL, E. Sex differences in adaptation of the GSR under repeated applications of a visual stimulus. *Journal of Experimental Psychology,* 1965, **70**, 536–537.

KIMMEL, H. D., & PENNYPACKER, H. S. Differential GSR conditioning as a function of the CS-UCS interval. *Journal of Experimental Psychology,* 1963, **65**, 559–563.

KIMMEL, H. D., & TERRANT, F. R. Bias due to individual differences in yoked control designs. *Behavior Research Methods and Instrumentation*, 1968, **1**, 11–14.

KOHLER, W. *Gestalt psychology*. New York: Liveright, 1947.

LANDSMAN, M. J. Perceptual accuracy as a function of stimulus value. Unpublished doctoral dissertation. University of Florida, 1964.

MAIER, N. R. F. *Frustration: The study of behavior without a good*. New York: McGraw-Hill, 1949.

MCGEOCH, J. A., & IRION, A. L. *The psychology of human learning*. New York: Longmans, 1952.

MCNEMAR, Q. *Psychological statistics*. New York: Wiley, 1955.

MEEHL, P. E. Theory-testing in psychology and physics: A methodological paradox. *Philosophy of Science*, 1967, **34**, 103–115.

MELVIN, K. B. Escape learning and vicious-circle behavior as a function of percentage of reinforcement. *Journal of Comparative and Physiological Psychology*, 1964, **58**, 248–251.

MORROW, M. C. Recovery of conditioned UCR diminution following extinction. *Journal of Experimental Psychology*, 1966, **71**, 884–888.

MOSTELLER, F. Remarks on the method of paired comparisons: III. A test of significance for paired comparisons when equal standard deviations and equal correlations are assumed. *Psychometrika*, 1951, **16**, 375–382.

OLDS, J., & MILNER, P. Positive reinforcement produced by electrical stimulation of septal area and other regions of rat brain. *Journal of Comparative and Physiological Psychology*, 1954, **47**, 419–427.

PAVLOV, I. P. *Conditioned reflexes* (Tr. & Ed. by G. V. Anrep). New York: Dover Publications, 1960.

PENDERGRASS, V. E., & KIMMEL, H. D. UCR diminution in temporal conditioning and habituation. *Journal of Experimental Psychology*, 1968, **77**, 1–6.

POSTMAN, L. The experimental analysis of motivational factors in perception. In M. R. Jones (Ed.), *Current theory and research in motivation*. Lincoln: University of Nebraska Press, 1953.

RAZRAN, G. Evolutionary psychology: Levels of learning–and perception and thinking. In B. B. Wolman (Ed.), *Scientific psychology*, New York: Basic Books, 1965.

RESCORLA, R. A., & SOLOMON, R. L. Two-process learning theory: Relationships between Pavlovian conditioning and instrumental learning. *Psychological Review*, 1967, **74**, 151–182.

RYAN, T. A. Multiple comparisons in psychological research. *Psychological Bulletin*, 1959, **56**, 26–47.

SANDLER, J. Resistance to punished extinction and amount of prior avoidance training. In H. D. Kimmel (Ed.), *Experimental Psychopathology: Research and Theory*. New York: Academic Press, 1971.

SCHULTZ, E. F., JR. Rules of thumb for determining expectations of mean squares in analysis of variance. *Biometrics*, 1958, **5**, 123–135.

SENDERS, V. L. *Measurement and statistics*. New York: Oxford University Press, 1958.

SIDMAN, M. Avoidance conditioning with brief shock and no exteroceptive warning signal. *Science*, 1953, **118**, 157–158.

SIDMAN, M. *Tactics of scientific research*. New York: Basic Books, 1960.

SIEGEL, S. *Nonparametric statistics*. New York: McGraw-Hill, 1956.

SKINNER, B. F. *Science and human behavior*. New York: Macmillan, 1953.

SKINNER, B. F. A case history in scientific method. *American Psychologist*, 1956, **11**, 221–223.

SKINNER, B. F. Teaching science in high school–What is wrong? *Science*, 1968, **159**, 704–710.

SPENCE, K. W. *Behavior theory and conditioning*. New Haven: Yale University Press, 1956.

STEVENS, S. S. On the psychophysical law. *Psychological Review,* 1957, **64**, 153–181.

STEVENS, S. S. The psychophysics of sensory function. In W. A. Rosenbleth (Ed.), *Sensory communication.* Boston: MIT Press, 1961.

TAYLOR, D. W., & FAUST, W. L. Twenty questions: Efficiency in problem solving as a function of group size. *Journal of Experimental Psychology,* 1952, **44**, 360–368.

TAYLOR, J. A. The relationship of anxiety to the conditioned eyelid response. *Journal of Experimental Psychology,* 1951, **41**, 81–92.

THURSTONE, L. L. A law of comparative judgment. *Psychological Review,* 1927, **34**, 273–286.

TIMMONS, E. O. *Implications of awareness in applied areas.* Unpublished symposium paper (Southeastern Psychological Association), Miami, Fla. April, 1963.

TOLMAN, E. C. Principles of purposive behavior. In S. Koch (Ed.), *Psychology: A study of a science.* New York: McGraw-Hill, 1959.

TORGERSON, W. S. *Theory and methods of scaling.* New York: Wiley, 1958.

TUKEY, J. W. Analyzing data: Sanctification or detective work. *American Psychologist,* 1969, **24**, 83–91.

UNDERWOOD, B. J. *Psychological research.* New York: Appleton-Century, 1957.

WEBB, E. J., CAMPBELL, D. T., SCHWARTZ, R. D., & SECHREST, L. *Unobtrusive measures: Nonreactive research in the social sciences.* Chicago: Rand McNally, 1966.

WEINER, H. Conditioning history and human fixed-interval performance. *Journal of the Experimental Analysis of Behavior,* 1964, **7**, 383–385.

WEINER, H. Controlling human fixed-interval performance. *Journal of the Experimental Analysis of Behavior,* 1969, **12**, 349–373.

WILCOXON, H. C. "Abnormal fixation" and learning. *Journal of Experimental Psychology,* 1952, **44**, 324–333.

WINER, B. J. *Statistical principles in experimental design.* New York: McGraw-Hill, 1962.

YAREMKO, R. M. The effects of direction and amount of change in stimulus intensity on magnitude of the orienting reflex and frequency of signal detection. Unpublished doctoral dissertation, Ohio University, 1969.

ZEIGARNIK, B. Über das Behalten erledigten und unerledigten Handlungen. *Psychologische Forschung,* 1927, **9**, 1–85.

Appendix A

The cumulative normal distribution function: cumulative proportions to unit normal deviates.

(for $P < 0.500$, use $1.000 - P$ and find $X_{1\text{-}P} = -X_P$)

P	0.000	0.001	0.002	0.003	0.004	0.005	0.006	0.007	0.008	0.009
0.50	0.000	0.003	0.005	0.008	0.010	0.013	0.015	0.018	0.020	0.023
0.51	0.025	0.028	0.030	0.033	0.035	0.038	0.040	0.043	0.045	0.048
0.52	0.050	0.053	0.055	0.058	0.060	0.063	0.065	0.068	0.070	0.073
0.53	0.075	0.078	0.080	0.083	0.085	0.088	0.090	0.093	0.095	0.098
0.54	0.100	0.103	0.105	0.108	0.111	0.113	0.116	0.118	0.121	0.123
0.55	0.126	0.128	0.131	0.133	0.136	0.138	0.141	0.143	0.146	0.148
0.56	0.151	0.154	0.156	0.159	0.161	0.164	0.166	0.169	0.171	0.174
0.57	0.176	0.179	0.181	0.184	0.187	0.189	0.192	0.194	0.197	0.199
0.58	0.202	0.204	0.207	0.210	0.212	0.215	0.217	0.220	0.222	0.225
0.59	0.228	0.230	0.233	0.235	0.238	0.240	0.243	0.246	0.248	0.251
0.60	0.253	0.256	0.259	0.261	0.264	0.266	0.269	0.272	0.274	0.277
0.61	0.279	0.282	0.285	0.287	0.290	0.292	0.295	0.298	0.300	0.303
0.62	0.305	0.308	0.311	0.313	0.316	0.319	0.321	0.324	0.327	0.329
0.63	0.332	0.335	0.337	0.340	0.342	0.345	0.348	0.350	0.353	0.356
0.64	0.358	0.361	0.364	0.366	0.369	0.372	0.375	0.377	0.380	0.383
0.65	0.385	0.388	0.391	0.393	0.396	0.399	0.402	0.404	0.407	0.410
0.66	0.412	0.415	0.418	0.421	0.423	0.426	0.429	0.432	0.434	0.437
0.67	0.440	0.443	0.445	0.448	0.451	0.454	0.457	0.459	0.462	0.465
0.68	0.468	0.470	0.473	0.476	0.479	0.482	0.485	0.487	0.490	0.493
0.69	0.496	0.499	0.502	0.504	0.507	0.510	0.513	0.516	0.519	0.522
0.70	0.524	0.527	0.530	0.533	0.536	0.539	0.542	0.545	0.548	0.550
0.71	0.553	0.556	0.559	0.562	0.565	0.568	0.571	0.574	0.577	0.580
0.72	0.583	0.586	0.589	0.592	0.595	0.598	0.601	0.604	0.607	0.610
0.73	0.613	0.616	0.619	0.622	0.625	0.628	0.631	0.634	0.637	0.640
0.74	0.643	0.646	0.650	0.653	0.656	0.659	0.662	0.665	0.668	0.671
0.75	0.674	0.678	0.681	0.684	0.687	0.690	0.693	0.697	0.700	0.703
0.76	0.706	0.710	0.713	0.716	0.719	0.722	0.726	0.729	0.732	0.736
0.77	0.739	0.742	0.745	0.749	0.752	0.755	0.759	0.762	0.765	0.769
0.78	0.772	0.776	0.779	0.782	0.786	0.789	0.793	0.796	0.800	0.803
0.79	0.806	0.810	0.813	0.817	0.820	0.824	0.827	0.831	0.834	0.838
0.80	0.842	0.845	0.849	0.852	0.856	0.860	0.863	0.867	0.871	0.874
0.81	0.878	0.882	0.885	0.889	0.893	0.896	0.900	0.904	0.908	0.912
0.82	0.915	0.919	0.923	0.927	0.931	0.935	0.938	0.942	0.946	0.950
0.83	0.954	0.958	0.962	0.966	0.970	0.974	0.978	0.982	0.986	0.990
0.84	0.994	0.999	1.003	1.007	1.011	1.015	1.019	1.024	1.028	1.032
0.85	1.036	1.041	1.045	1.049	1.054	1.058	1.063	1.067	1.071	1.076
0.86	1.080	1.085	1.089	1.094	1.098	1.103	1.108	1.112	1.117	1.122
0.87	1.126	1.131	1.136	1.141	1.146	1.150	1.155	1.160	1.165	1.170
0.88	1.175	1.180	1.185	1.190	1.195	1.200	1.206	1.211	1.216	1.221
0.89	1.227	1.232	1.237	1.243	1.248	1.254	1.259	1.265	1.270	1.276
0.90	1.282	1.287	1.293	1.299	1.305	1.311	1.317	1.323	1.329	1.335
0.91	1.341	1.347	1.353	1.359	1.366	1.372	1.379	1.385	1.392	1.398
0.92	1.405	1.412	1.419	1.426	1.433	1.440	1.447	1.454	1.461	1.468
0.93	1.476	1.483	1.491	1.499	1.506	1.514	1.522	1.530	1.538	1.546
0.94	1.555	1.563	1.572	1.580	1.589	1.598	1.607	1.616	1.626	1.635
0.95	1.645	1.655	1.665	1.675	1.685	1.695	1.706	1.717	1.728	1.739
0.96	1.751	1.762	1.774	1.787	1.799	1.812	1.825	1.838	1.852	1.866
0.97	1.881	1.896	1.911	1.927	1.943	1.960	1.977	1.995	2.014	2.034
0.98	2.054	2.075	2.097	2.120	2.144	2.170	2.197	2.226	2.257	2.290
0.99	2.326	2.366	2.409	2.457	2.512	2.576	2.652	2.748	2.878	3.090

Appendix B

Cumulative Normal Probability Function

	0.00	0.01	0.02	0.03	0.04	
0.0	0.500 00	0.503 99	0.507 98	0.511 97	0.515 95	**0.0**
0.1	0.539 83	0.543 79	0.547 76	0.551 72	0.555 67	**0.1**
0.2	0.579 26	0.583 17	0.587 06	0.590 95	0.594 83	**0.2**
0.3	0.617 91	0.621 72	0.625 51	0.629 30	0.633 07	**0.3**
0.4	0.655 42	0.659 10	0.662 76	0.666 40	0.670 03	**0.4**
0.5	0.691 46	0.694 97	0.698 47	0.701 94	0.705 40	**0.5**
0.6	0.725 75	0.729 07	0.732 37	0.735 65	0.738 91	**0.6**
0.7	0.758 03	0.761 15	0.764 24	0.767 30	0.770 35	**0.7**
0.8	0.788 14	0.791 03	0.793 89	0.796 73	0.799 54	**0.8**
0.9	0.815 94	0.818 59	0.821 21	0.823 81	0.826 39	**0.9**
1.0	0.841 34	0.843 75	0.846 13	0.848 49	0.850 83	**1.0**
1.1	0.864 33	0.866 50	0.868 64	0.870 76	0.872 85	**1.1**
1.2	0.884 93	0.886 86	0.888 77	0.890 65	0.892 51	**1.2**
1.3	0.903 20	0.904 90	0.906 58	0.908 24	0.909 88	**1.3**
1.4	0.919 24	0.920 73	0.922 19	0.923 64	0.925 06	**1.4**
1.5	0.933 19	0.934 48	0.935 74	0.936 99	0.938 22	**1.5**
1.6	0.945 20	0.946 30	0.947 38	0.948 45	0.949 50	**1.6**
1.7	0.955 43	0.956 37	0.957 28	0.958 18	0.959 07	**1.7**
1.8	0.964 07	0.964 85	0.965 62	0.966 37	0.967 11	**1.8**
1.9	0.971 28	0.971 93	0.972 57	0.973 20	0.973 81	**1.9**
2.0	0.977 25	0.977 78	0.978 31	0.978 82	0.979 32	**2.0**
2.1	0.982 14	0.982 57	0.983 00	0.983 41	0.983 82	**2.1**
2.2	0.986 10	0.986 45	0.986 79	0.987 13	0.987 45	**2.2**
2.3	0.989 28	0.989 56	0.989 83	0.990 10	0.990 36	**2.3**
2.4	0.991 80	0.992 02	0.992 24	0.992 45	0.992 66	**2.4**
2.5	0.993 79	0.993 96	0.994 13	0.994 30	0.994 46	**2.5**
2.6	0.995 34	0.995 47	0.995 60	0.995 73	0.995 85	**2.6**
2.7	0.996 53	0.996 64	0.996 74	0.996 83	0.996 93	**2.7**
2.8	0.997 44	0.997 52	0.997 60	0.997 67	0.997 74	**2.8**
2.9	0.998 13	0.998 19	0.998 25	0.998 31	0.998 36	**2.9**
3.0	0.998 65	0.998 69	0.998 74	0.998 78	0.998 82	**3.0**
3.1	0.999 03	0.999 06	0.999 10	0.999 13	0.999 16	**3.1**
3.2	0.999 31	0.999 34	0.999 36	0.999 38	0.999 40	**3.2**
3.3	0.999 52	0.999 53	0.999 55	0.999 57	0.999 58	**3.3**
3.4	0.999 66	0.999 68	0.999 69	0.999 70	0.999 71	**3.4**
3.5	0.999 77	0.999 78	0.999 78	0.999 79	0.999 80	**3.5**
3.6	0.999 84	0.999 85	0.999 85	0.999 86	0.999 86	**3.6**
3.7	0.999 89	0.999 90	0.999 90	0.999 90	0.999 91	**3.7**
3.8	0.999 93	0.999 93	0.999 93	0.999 94	0.999 94	**3.8**
3.9	0.999 95	0.999 95	0.999 96	0.999 96	0.999 96	**3.9**

APPENDIX B

Cumulative Normal Probability Function

	0.05	0.06	0.07	0.08	0.09	
0.0	0.519 94	0.523 92	0.527 90	0.531 88	0.535 86	**0.0**
0.1	0.559 62	0.563 56	0.567 49	0.571 42	0.575 34	**0.1**
0.2	0.598 71	0.602 57	0.606 42	0.610 26	0.614 09	**0.2**
0.3	0.636 83	0.640 58	0.644 31	0.648 03	0.651 73	**0.3**
0.4	0.673 64	0.677 24	0.680 82	0.684 38	0.687 93	**0.4**
0.5	0.708 84	0.712 26	0.715 66	0.719 04	0.722 40	**0.5**
0.6	0.742 15	0.745 37	0.748 57	0.751 75	0.754 90	**0.6**
0.7	0.773 37	0.776 37	0.779 35	0.782 30	0.785 23	**0.7**
0.8	0.802 34	0.805 10	0.807 85	0.810 57	0.813 27	**0.8**
0.9	0.828 94	0.831 47	0.833 97	0.836 46	0.838 91	**0.9**
1.0	0.853 14	0.855 43	0.857 69	0.859 93	0.862 14	**1.0**
1.1	0.874 93	0.876 97	0.879 00	0.881 00	0.882 97	**1.1**
1.2	0.894 35	0.896 16	0.897 96	0.899 73	0.901 47	**1.2**
1.3	0.911 49	0.913 08	0.914 65	0.916 21	0.917 73	**1.3**
1.4	0.926 47	0.927 85	0.929 22	0.930 56	0.931 89	**1.4**
1.5	0.939 43	0.940 62	0.941 79	0.942 95	0.944 08	**1.5**
1.6	0.950 53	0.951 54	0.952 54	0.953 52	0.954 48	**1.6**
1.7	0.959 94	0.960 80	0.961 64	0.962 46	0.963 27	**1.7**
1.8	0.967 84	0.968 56	0.969 26	0.969 95	0.970 62	**1.8**
1.9	0.974 41	0.975 00	0.975 58	0.976 15	0.976 70	**1.9**
2.0	0.979 82	0.980 30	0.980 77	0.981 24	0.981 69	**2.0**
2.1	0.984 22	0.984 61	0.985 00	0.985 37	0.985 74	**2.1**
2.2	0.987 78	0.988 09	0.988 40	0.988 70	0.988 99	**2.2**
2.3	0.990 61	0.990 86	0.991 11	0.991 34	0.991 58	**2.3**
2.4	0.992 86	0.993 05	0.993 24	0.993 43	0.993 61	**2.4**
2.5	0.994 61	0.994 77	0.994 92	0.995 06	0.995 20	**2.5**
2.6	0.995 98	0.996 09	0.996 21	0.996 32	0.996 43	**2.6**
2.7	0.997 02	0.997 11	0.997 20	0.997 28	0.997 36	**2.7**
2.8	0.997 81	0.997 88	0.997 95	0.998 01	0.998 07	**2.8**
2.9	0.998 41	0.998 46	0.998 51	0.998 56	0.998 61	**2.9**
3.0	0.998 86	0.998 89	0.998 93	0.998 97	0.999 00	**3.0**
3.1	0.999 18	0.999 21	0.999 24	0.999 26	0.999 29	**3.1**
3.2	0.999 42	0.999 44	0.999 46	0.999 48	0.999 50	**3.2**
3.3	0.999 60	0.999 61	0.999 62	0.999 64	0.999 65	**3.3**
3.4	0.999 72	0.999 73	0.999 74	0.999 75	0.999 76	**3.4**
3.5	0.999 81	0.999 81	0.999 82	9.999 83	0.999 83	**3.5**
3.6	0.999 87	0.999 87	0.999 88	0.999 88	0.999 89	**3.6**
3.7	0.999 91	0.999 92	0.999 92	0.999 92	0.999 92	**3.7**
3.8	0.999 94	0.999 94	0.999 95	0.999 95	0.999 95	**3.8**
3.9	0.999 96	0.999 96	0.999 96	0.999 97	0.999 97	**3.9**

Appendix C

Table of the *F* Distribution

$P = 0.01$
Values of *F* for right tail of the distribution

n_1/n_2	1	2	3	4	5	6	7	8	9
1	4052.	4999.	5403.	5625.	5764.	5859.	5928.	5982.	6022.
2	98.50	99.00	99.17	99.25	99.30	99.33	99.36	99.37	99.39
3	34.12	30.82	29.46	28.71	28.24	27.91	27.67	27.49	27.34
4	21.20	18.00	16.69	15.98	15.52	15.21	14.98	14.80	14.66
5	16.26	13.27	12.06	11.39	10.97	10.67	10.46	10.29	10.16
6	13.74	10.92	9.78	9.15	8.75	8.47	8.26	8.10	7.98
7	12.25	9.55	8.45	7.85	7.46	7.19	6.99	6.84	6.72
8	11.26	8.65	7.59	7.01	6.63	6.37	6.18	6.03	5.91
9	10.56	8.02	6.99	6.42	6.06	5.80	5.61	5.47	5.35
10	10.04	7.56	6.55	5.99	5.64	5.39	5.20	5.06	4.94
11	9.65	7.21	6.22	5.67	5.32	5.07	4.89	4.74	4.63
12	9.33	6.93	5.95	5.41	5.06	4.82	4.64	4.50	4.39
13	9.07	6.70	5.74	5.21	4.86	4.62	4.44	4.30	4.19
14	8.86	6.51	5.56	5.04	4.69	4.46	4.28	4.14	4.03
15	8.68	6.36	5.42	4.89	4.56	4.32	4.14	4.00	3.89
16	8.53	6.23	5.29	4.77	4.44	4.20	4.03	3.89	3.78
17	8.40	6.11	5.18	4.67	4.34	4.10	3.93	3.79	3.68
18	8.29	6.01	5.09	4.58	4.25	4.01	3.84	3.71	3.60
19	8.18	5.93	5.01	4.50	4.17	3.94	3.77	3.63	3.52
20	8.10	5.85	4.94	4.43	4.10	3.87	3.70	3.56	3.46
21	8.02	5.78	4.87	4.37	4.04	3.81	3.64	3.51	3.40
22	7.95	5.72	4.82	4.31	3.99	3.76	3.59	3.45	3.35
23	7.88	5.66	4.76	4.26	3.94	3.71	3.54	3.41	3.30
24	7.82	5.61	4.72	4.22	3.90	3.67	3.50	3.36	3.26
25	7.77	5.57	4.68	4.18	3.85	3.63	3.46	3.32	3.22
26	7.72	5.53	4.64	4.14	3.82	3.59	3.42	3.29	3.18
27	7.68	5.49	4.60	4.11	3.78	3.56	3.39	3.26	3.15
28	7.64	5.45	4.57	4.07	3.75	3.53	3.36	3.23	3.12
29	7.60	5.42	4.54	4.04	3.73	3.50	3.33	3.20	3.09
30	7.56	5.39	4.51	4.02	3.70	3.47	3.30	3.17	3.07
40	7.31	5.18	4.31	3.83	3.51	3.29	3.12	2.99	2.89
60	7.08	4.98	4.13	3.65	3.34	3.12	2.95	2.82	2.72
120	6.85	4.79	3.95	3.48	3.17	2.96	2.79	2.66	2.56
∞	6.63	4.61	3.78	3.32	3.02	2.80	2.64	2.51	2.41

Table of the *F* Distribution

$P = 0.01$

Values of *F* for right tail of the distribution

10	15	20	24	30	40	60	120	∞	n_1/n_2
6056.	6157.	6209.	6235.	6261.	6287.	6313.	6339.	6366.	1
99.40	99.43	99.45	99.46	99.47	99.47	99.48	99.49	99.50	2
27.23	26.87	26.69	26.60	26.50	26.41	26.32	26.22	26.12	3
14.55	14.20	14.02	13.93	13.84	13.74	13.65	13.56	13.46	4
10.05	9.72	9.55	9.47	9.38	9.29	9.20	9.11	9.02	5
7.87	7.56	7.40	7.31	7.23	7.14	7.06	6.97	6.88	6
6.62	6.31	6.16	6.07	5.99	5.91	5.82	5.74	5.65	7
5.81	5.52	5.36	5.28	5.20	5.12	5.03	4.95	4.86	8
5.26	4.96	4.81	4.73	4.65	4.57	4.48	4.40	4.31	9
4.85	4.56	4.41	4.33	4.25	4.17	4.08	4.00	3.91	10
4.54	4.25	4.10	4.02	3.94	3.86	3.78	3.69	3.60	11
4.30	4.01	3.86	3.78	3.70	3.62	3.54	3.45	3.36	12
4.10	3.82	3.66	3.59	3.51	3.43	3.34	3.25	3.17	13
3.94	3.66	3.51	3.43	3.35	3.27	3.18	3.09	3.00	14
3.80	3.52	3.37	3.29	3.21	3.13	3.05	2.96	2.87	15
3.69	3.41	3.26	3.18	3.10	3.02	2.93	2.84	2.75	16
3.59	3.31	3.16	3.08	3.00	2.92	2.83	2.75	2.65	17
3.51	3.23	3.08	3.00	2.92	2.84	2.75	2.66	2.57	18
3.43	3.15	3.00	2.92	2.84	2.76	2.67	2.58	2.49	19
3.37	3.09	2.94	2.86	2.78	2.69	2.61	2.52	2.42	20
3.31	3.03	2.88	2.80	2.72	2.64	2.55	2.46	2.36	21
3.26	2.98	2.83	2.75	2.67	2.58	2.50	2.40	2.31	22
3.21	2.93	2.78	2.70	2.62	2.54	2.45	2.35	2.26	23
3.17	2.89	2.74	2.66	2.58	2.49	2.40	2.31	2.21	24
3.13	2.85	2.70	2.62	2.54	2.45	2.36	2.27	2.17	25
3.09	2.81	2.66	2.58	2.50	2.42	2.33	2.23	2.13	26
3.06	2.78	2.63	2.55	2.47	2.38	2.29	2.20	2.10	27
3.03	2.75	2.60	2.52	2.44	2.35	2.26	2.17	2.06	28
3.00	2.73	2.57	2.49	2.41	2.33	2.23	2.14	2.03	29
2.98	2.70	2.55	2.47	2.39	2.30	2.21	2.11	2.01	30
2.80	2.52	2.37	2.29	2.20	2.11	2.02	1.92	1.80	40
2.63	2.35	2.20	2.12	2.03	1.94	1.84	1.73	1.60	60
2.47	2.19	2.03	1.95	1.86	1.76	1.66	1.53	1.38	120
2.32	2.04	1.88	1.79	1.70	1.59	1.47	1.32	1.00	∞

Table of the *F* Distribution

$P = 0.05$

Values of *F* for right tail of the distribution

n_1/n_2	1	2	3	4	5	6	7	8	9
1	161.45	199.50	215.71	224.58	230.16	233.99	236.77	238.88	240.54
2	18.51	19.00	19.16	19.25	19.30	19.33	19.35	19.37	19.38
3	10.13	9.55	9.28	9.12	9.01	8.94	8.89	8.85	8.81
4	7.71	6.94	6.59	6.39	6.26	6.16	6.09	6.04	6.00
5	6.61	5.79	5.41	5.19	5.05	4.95	4.88	4.82	4.77
6	5.99	5.14	4.76	4.53	4.39	4.28	4.21	4.15	4.10
7	5.59	4.74	4.35	4.12	3.97	3.87	3.79	3.73	3.68
8	5.32	4.46	4.07	3.84	3.69	3.58	3.50	3.44	3.39
9	5.12	4.26	3.86	3.63	3.48	3.37	3.29	3.23	3.18
10	4.96	4.10	3.71	3.48	3.33	3.22	3.14	3.07	3.02
11	4.84	3.98	3.59	3.36	3.20	3.09	3.01	2.95	2.90
12	4.75	3.89	3.49	3.26	3.11	3.00	2.91	2.85	2.80
13	4.67	3.81	3.41	3.18	3.03	2.92	2.83	2.77	2.71
14	4.60	3.74	3.34	3.11	2.96	2.85	2.76	2.70	2.65
15	4.54	3.68	3.29	3.06	2.90	2.79	2.71	2.64	2.59
16	4.49	3.63	3.24	3.01	2.85	2.74	2.66	2.59	2.54
17	4.45	3.59	3.20	2.96	2.81	2.70	2.61	2.55	2.49
18	4.41	3.55	3.16	2.93	2.77	2.66	2.58	2.51	2.46
19	4.38	3.52	3.13	2.90	2.74	2.63	2.54	2.48	2.42
20	4.35	3.49	3.10	2.87	2.71	2.60	2.51	2.45	2.39
21	4.32	3.47	3.07	2.84	2.68	2.57	2.49	2.42	2.37
22	4.30	3.44	3.05	2.82	2.66	2.55	2.46	2.40	2.34
23	4.28	3.42	3.03	2.80	2.64	2.53	2.44	2.37	2.32
24	4.26	3.40	3.01	2.78	2.62	2.51	2.42	2.36	2.30
25	4.24	3.39	2.99	2.76	2.60	2.49	2.40	2.34	2.28
26	4.23	3.37	2.98	2.74	2.59	2.47	2.39	2.32	2.27
27	4.21	3.35	2.96	2.73	2.57	2.46	2.37	2.31	2.25
28	4.20	3.34	2.95	2.71	2.56	2.45	2.36	2.29	2.24
29	4.18	3.33	2.93	2.70	2.55	2.43	2.35	2.28	2.22
30	4.17	3.32	2.92	2.69	2.53	2.42	2.33	2.27	2.21
40	4.08	3.23	2.84	2.61	2.45	2.34	2.25	2.18	2.12
60	4.00	3.15	2.76	2.53	2.37	2.25	2.17	2.10	2.04
120	3.92	3.07	2.68	2.45	2.29	2.17	2.09	2.02	1.96
∞	3.84	3.00	2.60	2.37	2.21	2.10	2.01	1.94	1.88

Table of the *F* Distribution

$P = 0.05$

Values of *F* for right tail of the distribution

10	15	20	24	30	40	60	120	∞	n_1 / n_2
241.88	245.95	248.01	249.05	250.09	251.14	252.20	253.25	254.32	1
19.40	19.43	19.45	19.45	19.46	19.47	19.48	19.49	19.50	2
8.76	8.70	8.66	8.64	8.62	8.59	8.57	8.55	8.53	3
5.96	5.86	5.80	5.77	5.75	5.72	5.69	5.66	5.63	4
4.74	4.62	4.56	4.53	4.50	4.46	4.43	4.40	4.36	5
4.06	3.94	3.87	3.84	3.81	3.77	3.74	3.70	3.67	6
3.64	3.51	3.44	3.41	3.38	3.34	3.30	3.27	3.23	7
3.35	3.22	3.15	3.12	3.08	3.04	3.01	2.97	2.93	8
3.14	3.01	2.94	2.90	2.86	2.83	2.79	2.75	2.71	9
2.98	2.84	2.77	2.74	2.70	2.66	2.62	2.58	2.54	10
2.85	2.72	2.65	2.61	2.57	2.53	2.49	2.45	2.40	11
2.75	2.62	2.54	2.51	2.47	2.43	2.38	2.34	2.30	12
2.67	2.53	2.46	2.42	2.38	2.34	2.30	2.25	2.21	13
2.60	2.46	2.39	2.35	2.31	2.27	2.22	2.18	2.13	14
2.54	2.40	2.33	2.29	2.25	2.20	2.16	2.11	2.07	15
2.49	2.35	2.28	2.24	2.19	2.15	2.11	2.06	2.01	16
2.45	2.31	2.23	2.19	2.15	2.10	2.06	2.01	1.96	17
2.41	2.27	2.19	2.15	2.11	2.06	2.02	1.97	1.92	18
2.38	2.23	2.16	2.11	2.07	2.03	1.98	1.93	1.88	19
2.35	2.20	2.12	2.08	2.04	1.99	1.95	1.90	1.84	20
2.32	2.18	2.10	2.05	2.01	1.96	1.92	1.87	1.81	21
2.30	2.15	2.07	2.03	1.98	1.94	1.89	1.84	1.78	22
2.27	2.13	2.05	2.00	1.96	1.91	1.86	1.81	1.76	23
2.25	2.11	2.03	1.98	1.94	1.89	1.84	1.79	1.73	24
2.24	2.09	2.01	1.96	1.92	1.87	1.82	1.77	1.71	25
2.22	2.07	1.99	1.95	1.90	1.85	1.80	1.75	1.69	26
2.20	2.06	1.97	1.93	1.88	1.84	1.79	1.73	1.67	27
2.19	2.04	1.96	1.91	1.87	1.82	1.77	1.71	1.65	28
2.18	2.03	1.94	1.90	1.85	1.81	1.75	1.70	1.64	29
2.16	2.01	1.93	1.89	1.84	1.79	1.74	1.68	1.62	30
2.08	1.92	1.84	1.79	1.74	1.69	1.64	1.58	1.51	40
1.99	1.84	1.75	1.70	1.65	1.59	1.53	1.47	1.39	60
1.91	1.75	1.66	1.61	1.55	1.50	1.43	1.35	1.25	120
1.83	1.67	1.57	1.52	1.46	1.39	1.31	1.22	1.00	∞

Appendix D

Confidence Factor *t**

					P						
DF	**0.90**	**0.80**	**0.70**	**0.60**	**0.50**	**0.25**	**0.10**	**0.05**	**0.025**	**0.010**	**0.005**
1	0.158	0.325	0.510	0.727	1.000	2.414	6.314	12.71	25.45	63.66	127.3
2	0.142	0.289	0.445	0.617	0.817	1.604	2.920	4.303	6.205	9.925	14.09
3	0.137	0.277	0.424	0.584	0.765	1.423	2.353	3.183	4.177	5.841	7.453
4	0.134	0.271	0.414	0.569	0.741	1.344	2.132	2.776	3.495	4.604	5.598
5	0.132	0.267	0.408	0.559	0.727	1.301	2.015	2.571	3.163	4.032	4.773
6	0.131	0.265	0.404	0.553	0.718	1.273	1.943	2.447	2.969	3.707	4.317
7	0.130	0.263	0.402	0.549	0.711	1.254	1.895	2.365	2.841	3.500	4.029
8	0.130	0.262	0.399	0.546	0.706	1.240	1.860	2.306	2.752	3.355	3.833
9	0.129	0.261	0.398	0.543	0.703	1.230	1.833	2.262	2.685	3.250	3.690
10	0.129	0.260	0.397	0.542	0.700	1.221	1.813	2.228	2.634	3.169	3.581
11	0.129	0.260	0.396	0.540	0.697	1.215	1.796	2.201	2.593	3.106	3.500
12	0.128	0.259	0.395	0.539	0.695	1.209	1.782	2.179	2.560	3.055	3.428
13	0.128	0.259	0.394	0.538	0.694	1.204	1.771	2.160	2.533	3.012	3.373
14	0.128	0.258	0.393	0.537	0.692	1.200	1.761	2.145	2.510	2.977	3.326
15	0.128	0.258	0.393	0.536	0.691	1.197	1.753	2.132	2.490	2.947	3.286
20	0.127	0.257	0.391	0.533	0.687	1.185	1.725	2.086	2.423	2.845	3.153
25	0.127	0.256	0.390	0.531	0.684	1.178	1.708	2.060	2.385	2.787	3.078
30	0.127	0.256	0.389	0.530	0.683	1.173	1.697	2.042	2.360	2.750	3.030
40	0.126	0.255	0.388	0.529	0.681	1.167	1.684	2.021	2.329	2.705	2.971
60	0.126	0.254	0.387	0.527	0.679	1.162	1.671	2.000	2.299	2.660	2.915
120	0.126	0.254	0.386	0.526	0.677	1.156	1.658	1.980	2.270	2.617	2.860
∞	0.126	0.253	0.385	0.524	0.674	1.150	1.645	1.960	2.241	2.576	2.807

* Two-tailed probabilities *P* for given degrees of freedom, *DF*.

Appendix E

Coefficients for Obtaining the Linear and Quadratic Components of the Treatment Sum of Squares When Treatments are Equally Spaced

		Number of Treatments								
Σk^2	Comparison	1	2	3	4	5	6	7	8	9
2	LINEAR	−1	0	1						
6	QUADRATIC	1	−2	1						
20	LINEAR	−3	−1	1	3					
4	QUADRATIC	1	−1	−1	1					
10	LINEAR	−2	−1	0	1	2				
14	QUADRATIC	2	−1	−2	−1	2				
70	LINEAR	−5	−3	−1	1	3	5			
84	QUADRATIC	5	−1	−4	−4	−1	5			
28	LINEAR	−3	−2	−1	0	1	2	3		
84	QUADRATIC	5	0	−3	−4	−3	0	5		
168	LINEAR	−7	−5	−3	−1	1	3	5	7	
168	QUADRATIC	7	1	−3	−5	−5	−3	1	7	
60	LINEAR	−4	−3	−2	−1	0	1	2	3	4
2,772	QUADRATIC	28	7	−8	−17	−20	−17	−8	7	28

Index